The Nontoxic Home

Also by Debra Lynn Dadd

Nontoxic & Natural: How to Avoid Dangerous Everyday Products and Buy or Make Safe Ones

Nontoxic, Natural, & Earthwise: How to Protect Yourself from Harmful Products and Live in Harmony with the Earth

The Nontoxic Home

Protecting Yourself and Your Family
from Everyday Toxics and Health
Hazards

Debra Lynn Dadd

JEREMY P. TARCHER, INC.
Los Angeles
Distributed by St. Martin's Press
New York

Library of Congress Cataloging in Publication Data

Dadd, Debra Lynn.
  The nontoxic home.
  Bibliography: p. 197
  Includes index.
  1. Toxicology—Popular works. 2. Product safety.
3. Consumer education. I. Title. [DNLM:
1. Consumer Product Safety—popular works.
2. Poisoning—prevention & control—popular works.
3. Toxicology—popular works. QV 600 D121n]
RA1213.D334 1986   615.9   86-14364
ISBN 0-87477-401-2 (pbk.)

Jeremy P. Tarcher, Inc.
5858 Wilshire Blvd., Suite 200
Los Angeles, CA 90036

Manufactured in the United States of America
10 9 8 7 6 5 4 3 2

for Sara and Bradley

ACKNOWLEDGMENTS

I would like to thank three special people for their contributions to this book:

Janice Gallagher, Executive Editor at Jeremy P. Tarcher, for coming up with the idea and asking me to write this book.

My father, Robert C. Dadd, for providing a beautiful and inspirational place to work.

Suzanne Lipsett, my wonderful editor, for her insight and intelligence, for keeping after me to make this book more personal, for asking all the perfect questions, for adapting to my hurried schedule, and for always being patient.

I would also like to acknowledge the thousands of people who have supported and helped me in my work over the past six years. You took the time to talk with me, answer my letters, send me your catalogs, cheer me on, tell others about my work, and share with me. I am able to write this book because of you.

CONTENTS

CONTENTS

Introduction

In writing this book, I speak to you not as a doctor or a scientist, but as an educated consumer. You will benefit from what I have learned in six years of firsthand experience living in a nontoxic home, and from my investigation of the subject as a full-time profession.

Researching toxic substances in everyday products and safe substitutes has been of vital interest to me since 1980, when I was diagnosed as having a severe breakdown of my immune system caused, I was told, by a combination of my high-stress lifestyle and heavy chemical exposure. I was shocked, for I thought my chemical exposure was no different from anyone else's living in our modern world. I worked as a professional classical musician and lived an urban existence in a major metropolitan city. I had no occupational exposure to chemicals—the chemicals that disabled me were all in my home.

Because my illness caused violent reactions to all man-made products, I began to look for things I could use that did not contain toxic substances or *any* petrochemical derivatives. After much trial and error, I finally came up with a small list. It was a difficult task because most toxic substances generally found in the home are *hidden* in products and the law does not require that they be listed on the label. For example, a canned food might contain residues of pesticides and fungicides, residues of the detergent used to wash the food, traces of lead from the solder used to hold the can together, and many kinds of artificial additives that could have been used in any of the ingredients. All the law requires the manufacturer to list on the label are the ingredients and additives mixed together in the final stage of processing, and not the additives and contaminants in the ingredients themselves. Ferreting out this kind of information is no mean task of detection. Most of my information about toxic chemicals in products has come from materials written for poison control centers, toxicology books, federal regulations, and trade journals—not from product labels.

My list of products and alternative methods steadily grew, through personal experience and contributions from friends, until finally I self-published a small directory of safe brand-name products, local stores where they could be purchased, and mail-order sources for some hard-to-obtain items. Soon my mailbox was filled with orders from all over the country, as news of this stapled stack of photocopies spread by word of mouth. My health improved quickly—due primarily to my new nontoxic lifestyle—and within a few months I was totally free of my formerly disabling symptoms.

The media picked up on my success story and I began receiving orders for my directory from people who weren't ill, but who were concerned about the environment or the health of their children, or from those who were just concerned about the possible long-term health effects of these unknown substances. In 1984, an expanded directory of safe products was published by J. P. Tarcher, *Nontoxic & Natural: How to Avoid Dangerous Everyday Products and Buy or Make Safe Ones.*

Almost immediately I began to get letters—new information on mail-order sources not listed in the book, feedback on do-it-yourself formulas, new suggestions, requests for almost-impossible-to-find products, questions on the toxicity of products, and much more. I saw that having my book published was only the beginning, and that what was necessary was an ongoing publication that would provide current information and act as a clearinghouse for many contributors. Thus, my newsletter *Everything Natural* was born. In the first eight issues, it evolved from two pieces of photocopied paper to a mini-magazine printed with edible vegetable-based ink. And the interest in and availability of natural products has grown. In 1982 I had to search hard for a small list of nontoxic and natural products. In 1986, there are so many new products, I can hardly keep track of them!

The Nontoxic Home is a response to the question I am asked most frequently: "Where do I begin?" Instead of listing resources, I'll discuss toxic chemicals and other health hazards found everywhere in your home, and give you information that will help you determine which pose the greatest risk, and which are easiest (and least expensive!) to change. This book is full of helpful hints and suggestions that will lead you to nontoxic products you might not even know exist!

Making changes in your home can be difficult, I know, especially when the amount of information to consider threatens to become overwhelming. I sat down to write this book thinking I would simply list and give alternatives for the hundred most toxic household products. A worker at my local poison control center laughed when I told her what I was trying to do. I quickly discovered it was an impossible task. So many factors determine toxicity that to say one thing is more toxic than another is like saying New York-style pizza is better than Chicago-style.

The best we can do is assess the danger of each product individually, learn the risks, weigh the risks against the benefits (and the low-risk and risk-free alternatives), and decide whether or not we personally want to take these risks and to subject those we love to them. The world is increasingly filled with health hazards, and such decisions start to seem endless. The place to begin is where we have the greatest power to make changes: in our homes.

CHAPTER 1

Discovering How Your Home
May Be Toxic

This book is for all those who want to protect themselves and others in their households from unnecessary exposure to toxic chemicals and other health hazards. We've been hearing a lot about toxic contaminants—in our water supply, in our food, in our air. It is becoming increasingly clear that our health is integrally related to the toxicity of our environment, but while television newsreels and daily newspapers run stories about chemical spills, toxic waste, and industrial polluters, little attention is given to positive, immediate actions we as individuals can take right now.

Help is here. *The Nontoxic Home* will help you gain some perspective on the problem of toxics, and how it affects you *in your own home.* It will give you the information you need to make choices and changes for a toxic-free home environment. The changes aren't going to be radical, the choices aren't going to be difficult—but simple as it is, in the long run *and* the short run, *this book could save your life.*

Why should you care about household toxics? Simple. Because your greatest exposure to toxic substances is right in your own home. Your house or apartment is probably full of everyday products made from materials and substances that cause cancer, birth defects, and changes in genetic structure, and that weaken the immune system, leaving your body vulnerable to many kinds of diseases and infections.

A multitude of common symptoms can be related to exposure to household toxics—headaches, depression, even ordinary flu symptoms might not be flu at all, but a pesticide poisoning or a reaction to your furniture polish.

Sometimes it's hardest to see what's closest to you, what you see every day. But let's take a quick mental tour of the typical American home—*your* home, perhaps—where you consider yourself and other members of your household safe from the dangers of the outside world. Mark a check next to the products you find in each room to see how safe you *really* are from household health hazards:

LIVING ROOM

☐ alcoholic beverages
☐ artificial light
☐ ionization-type smoke detectors
☐ fireplace
☐ furniture and floor polish
☐ gas heater
☐ house plants
☐ kerosene heater
☐ particleboard furniture
☐ rug, carpet, and upholstery shampoo
☐ spot remover
☐ synthetic wall-to-wall carpet
☐ tobacco smoke
☐ urea-formaldehyde foam insulation
☐ wood stove

KITCHEN

☐ aluminum cookware
☐ ammonia and all-purpose cleaners
☐ artificial sweeteners
☐ asbestos-vinyl floor tiles
☐ chlorinated scouring powder

- [ ] dishwasher detergent
- [ ] dishwashing liquid
- [ ] drain cleaners
- [ ] dried fruits
- [ ] fish and seafood
- [ ] gas appliances
- [ ] insecticides
- [ ] microwave ovens
- [ ] no-stick cookware
- [ ] oven cleaner
- [ ] particleboard cabinets
- [ ] plastic clock
- [ ] plastic drinking glasses
- [ ] plastic food wrap and storage containers
- [ ] plastic lighting fixtures
- [ ] plastic telephone
- [ ] processed foods
- [ ] rat and mouse killers
- [ ] silver polish and other metal cleaners
- [ ] supermarket eggs
- [ ] supermarket meat
- [ ] supermarket milk
- [ ] supermarket produce
- [ ] vitamin and mineral supplements
- [ ] canned food
- [ ] tap water
- [ ] chocolate and coffee
- [ ] store-bought water in plastic bottle

BATHROOM

- [ ] aerosol hair spray
- [ ] air fresheners
- [ ] antiperspirants
- [ ] astringents

- [ ] bubble bath
- [ ] contact lenses
- [ ] cosmetics
- [ ] dandruff shampoo
- [ ] denture cleaners
- [ ] deodorant soap
- [ ] disposable diapers
- [ ] drugs and medications
- [ ] feminine deodorant spray
- [ ] feminine douches
- [ ] fluoride mouthwash
- [ ] fluoride toothpaste
- [ ] germ-killing disinfectants
- [ ] glass cleaner
- [ ] hair color
- [ ] hair-removal products
- [ ] hair styling mousse
- [ ] mold and mildew cleaners
- [ ] nail polish and nail-polish remover
- [ ] perfume and aftershave
- [ ] permanent waves
- [ ] scented toilet paper
- [ ] superabsorbent tampons
- [ ] talcum powder
- [ ] vinyl shower curtain

BEDROOM

- [ ] children's sleepwear
- [ ] contraceptives
- [ ] dry cleaning
- [ ] fireproofed synthetic mattresses
- [ ] mothballs
- [ ] no-iron bed linens
- [ ] nylon stockings

☐ permanent-press clothing
☐ plastic baby pants
☐ plastic baby toys
☐ plastic raincoat and umbrella
☐ vinyl shoes and handbags

HOME OFFICE

☐ home computers
☐ permanent-ink pens and markers
☐ rubber cement and glues
☐ typewriter correction fluid

LAUNDRY ROOM

☐ chlorine bleach
☐ do-it-yourself dyes
☐ fabric softeners
☐ laundry detergents
☐ spray starch

How did you do? Are there lots of check marks? *Every one of the items on this list poses a health hazard*—either by immediate poisoning, or by slowly breaking down your health over a long period of time—*and there is a safe, nontoxic alternative* for each of these products.

The average home today contains more chemicals than were found in a typical chemistry lab at the turn of the century. When professionals use chemicals in industrial settings, they are subject to strict health and safety codes; yet we are allowed to use these same chemicals at home, without guidance or restriction.

Still, I hear many people say "I'm not sick, why should I worry?" If you are a young adult in good health, who is emotionally well balanced and not under stress, if you do not drink, smoke, or take drugs, get adequate nutrition, exercise regularly, and get enough rest, you probably are in good shape to withstand the barrage of toxic substances that assault you daily. Even so, most household toxics are universally injurious and harmful even to otherwise healthy people. The long-term health effects of many chemicals are unknown, and we do not know the possible synergistic effects that occur when chemicals are combined in food, water, or air, or when the chemicals are in your body interacting with other chemicals.

Children are particularly vulnerable to the health effects of household toxics. For example, simply because of children's size and physiology, they are exposed to more indoor pollutants than adults standing nearby in the same room! Children inhale more air per-body-weight than adults, their respiratory rates are *ten times faster* than those of adults, and, since pollutants are generally heavier than air, they are at greater concentrations at the height of a child's nose than they are at the height of an adult's nose. Even when children are actually exposed to the same amounts of chemicals as adults, their bodies are much smaller, so, relatively speaking, the proportion of their chemical exposure to body weight is greater than ours. The same is true for foods. Because children eat more in relation to their body weight, they consume proportionally more food additives and pesticides. Because children are more active than adults, they probably drink more water in relation to their body weight, and so have a greater exposure to water pollutants. At the same time, their bodies are not as well equipped to process toxic chemicals, so their risk is increased more than ours by two factors.

But slow contamination is not our only concern for the children in the house. Every year, 5 to 10 million household poisonings are reported. Many are fatal, and most of the victims are children. These poisonings are the result of accidental ingestion of common household substances, products that, despite showing warning labels, are not kept out of children's reach. In most cases, the suffering is entirely unnecessary, for the majority of these products have safe, effective alternatives that can allow you to remove potentially dangerous substances from your home altogether.

If you are thinking about conceiving a child, or are already pregnant, consider the effects toxic chemicals might have on your unborn baby. Many common household chemicals are known to cause birth defects, but most have not even been tested yet for these effects. This applies to men, too. Certain chemicals can damage sperm, and damaged sperm results in a deformed fetus. Other chemicals cause genetic changes, which can lead to health problems and birth defects in future generations.

Also consider the other end of the age spectrum. If you are feeling the effects of passing years, or are responsible for the care of an aging parent, you know that as we grow older, our bodies begin to wear down and don't work as well as they used to. This is true for your tolerance to chemical exposures, too. A good way to maintain good health, and perhaps add a few years to your life, is to live nontoxically. All the long-lived people of the world live in pristine environments, untouched by technological progress.

When you have an illness of any type, even a minor cold, your tolerance for household pollutants decreases. In fact, if you or a family member gets sick often, it may be an indicator that your home is highly toxic. Many subtle poisonings show up as flu symptoms and often go undetected by doctors.

Even the amount of stress you are under affects your body's ability to process toxic chemicals, and in today's high-stress society, we all qualify, even the young and healthy.

While there are certainly some people who are more sensitive to household pollutants than others, it is important to remember that these substances are poisoning all of us, whether we feel the effects immediately or not. Some toxics poison instantly, others show their devastating effects only after years of everyday exposure. How many times have we said we would have done something to prevent a tragedy, "If only I had known. . . . "

My intent here is not to alarm or overwhelm you, but to point you in a direction toward constructive change. We are grappling with a new problem here, for which there is no precedent. We have reached the point at which we *must* be informed consumers—we can no longer rely on our instinct, upbringing, or personal taste to guide us as we shop. We must learn the risks, and how to minimize them. If we ignore the toxic problems in our world, they won't go away—they will only get worse.

Don't expect to change your whole house overnight; the very nature of change is a gradual process. It begins with a decision, a commitment to change. Next, you have to take action. Simply reading this book will not protect you from common health hazards. You will need to act on the information you find here.

Deciding to make the changes is the hardest part. Finding nontoxic products is easy and can be fun, and you'll love the simplicity of the safe alternatives you can make at home. Following the suggestions outlined in this book just may save your child from an accidental fatal poisoning, and keep you from getting cancer. I know it seems like such a small thing to do, given the enormity of the problem of toxics in our world, but it's your *home,* your oasis of warmth and safety, where you alone are responsible for what you choose to put in it. There is a slogan environmentalists use, "Think globally, act locally." And remember that old saying, "Charity begins at home"? That's just where we have to begin with toxics, too.

HOW TO USE THIS BOOK

The arrangement of the chapters and of the items within them reflects the relative toxicity of the items discussed, with the most immediate hazards first. The exception is cigarette smoke, which, although most dangerous in the short and long run, is not as prevalent in our homes—we hope—as the other hazards. It happens that the products that are simplest and most inexpensive to change are also the most important to change first, as they are well-recognized poisons and major contributors to the toxicity of your home. If you are already aware of the dangers of food additives, you may be surprised to learn they are actually low on the list, after cleaning products, pesticides, and other household items. It actually may be more harmful to drink a glass of tap water than it is to eat processed foods!

While many, many alternative products and ideas might be included in a nontoxic home, I suggested those that I have found from personal experience to be the easiest and most effective. Also, this book does not cover *every* toxic product you might find in your home, only the most common—and the most dangerous.

Throughout the book I have included warnings from product labels when they appear. I call them Manufacturers' Warning Labels because they are compiled from an array of products. I thought it would be interesting to compare the label warnings with the *actual* dangers of the products, especially after my local poison control center told me that about *85 percent* of household items on the market are *mislabeled!* Some are products labeled poisons that really aren't, some are poisons not labeled as such, some labels warn of dangers but don't list the poison, and many contain incorrect first aid information. Also, label warnings are required only on products that are harmful or fatal if accidentally swallowed or inhaled in extreme concentrations. No warnings are given on products that have long-term health effects. In many cases, these effects are suspected, but currently unknown.

One very important factor to consider when assessing the risk of a product is how chemicals can enter your body. Some substances do not give off fumes, but can be deadly if swallowed. Others may cause skin irritation and also cause symptoms if they are inhaled. To help you quickly assess the danger, I have used four simple symbols to indicate by which modes of exposure the product may be harmful. In many cases there are secondary exposures that are not often recognized.

INGESTED BY MOUTH

SPLASHED IN THE EYES

ABSORBED THROUGH THE SKIN

**INHALED THROUGH
THE NOSE OR MOUTH**

Throughout the book I have emphasized the toxic substances found in products by having them set in boldface type. It's important that you begin to recognize the names of these dangerous chemicals and you will see that most are used in many products.

I have kept my comments simple, to give you an idea of what is possible and to inspire you to seek more information. It is almost impossible to be complete, because there is just so much information available, and much of it is beyond the scope of this book. For this reason, I did not include any brand-name products or mail-order sources of products in this book. If you would like information on specific sources for safe products, my first book, *Nontoxic & Natural: How to Avoid Dangerous Everyday Products and Buy or Make Safe Ones* and my bimonthly newsletter, *Everything Natural,* will be of immense help to you. Information on ordering these publications is in the Resources section at the end of the book, along with other sources of information.

DISPOSING OF HOUSEHOLD TOXICS

If you are like I was, once you're convinced of the dangers of certain products, you're going to want to throw everything away and start using nontoxic alternatives immediately. But when you do, you face the dilemma of proper disposal.

Why is disposal of household toxics a problem? Because we generate hazardous waste when we dispose of our household products. When we throw a half-used can of pesticide in the garbage can, it is taken to the city dump, and it may leach into our water supply, be sprayed on our croplands, and end up right back in our homes. According to an article in *National Geographic,* we throw away 4 million tons of hazardous waste each year. Disposal of household hazardous waste is becoming such a problem that many communities are setting up special disposal programs to keep their city refuse dumps from turning into toxic waste dumps.

Call your local public health department, environmental health department, department of health services, or sanitation department to see if you have a hazardous-waste disposal program in your community. They will generally accept such items as pesticides, household cleaners, paints and paint products, automotive products, solvents, pharmaceuticals, aerosol products, pool chemicals, hobby supplies, acids, and waste oil. If you don't have a local program, ask your local authorities how best to dispose of your toxics. It may be *illegal* to dispose of certain products with your normal household garbage.

If your local government has no suggestions, at least dilute small amounts of toxics with large amounts of water and dispose of them down the drain. If you have any doubts about what you can safely put down the drain, call your local sewer district. Your county department of agriculture may accept small amounts of pesticides, so call them and ask.

GENERAL PRECAUTIONS FOR USING HOUSEHOLD POISONS

The point of this book is to discourage you from using household toxics, but if you choose to continue using them, at least take some precautions to prevent accidental poisonings:

- Keep products in their original containers so the original label is available in case of accidental poisoning, and to prevent confusion. Accidental poisonings often occur when look-alike poisons are stored in food and drink containers.
- Don't use more of a product than the directions call for. Using more hardly ever makes a product more effective, and can be harmful.
- Keep containers tightly closed to prevent volatile fumes from escaping.
- Use products in well-ventilated areas, outdoors if possible, to avoid breathing fumes.
- Don't mix chemicals. Some combinations, such as chlorine bleach and ammonia (or products such as scouring powder and an all-purpose cleaner, which contain these substances), can produce toxic fumes. If you're not a chemist, you won't know what you'll end up with.
- Wear protective clothing if it is recommended on the label.

15

- Buy only what you need, and use it up, or dispose of it properly.
- Carefully clean up after using toxics. Make sure products are properly stored, spills are wiped up or washed away, and rags are properly disposed of.
- And, of course, keep out of reach of children.

EMERGENCY ACTION FOR HOUSEHOLD POISONINGS

If you follow the instructions in this book, you'll never need to know what to do in case of an accidental poisoning. But if you have toxics in the house, the American College of Emergency Physicians and the American Association of Poison Control Centers recommend the following emergency procedures:

"INHALED POISON—Immediately get the person to fresh air. Avoid breathing fumes. Open doors and windows wide. If victim is not breathing, start artificial respiration.

"POISON ON THE SKIN—Remove contaminated clothing and flood skin with water for 10 minutes. Then wash gently with soap and water and rinse.

"POISON IN THE EYE—Flood the eye with lukewarm (not hot) water poured from a large glass 2 or 3 inches from the eye. Repeat for 15 minutes. Have patient blink as much as possible while flooding the eye. Do not force the eyelid open.

"SWALLOWED POISON—Unless patient is unconscious, having convulsions, or cannot swallow—give milk or water immediately, then call for professional advice about whether you should make the patient vomit or not. ALWAYS KEEP ON HAND AT HOME a one-ounce bottle of SYRUP OF IPECAC for each child in the home. Use only on advice of poison control center, emergency department, or physicians."

Then, ignore instructions on the product label and call your local poison control center immediately! If you have toxics and small children in the house, keep that emergency telephone number next to the telephone and teach your children when they should use it and why.

CHAPTER 2

Cleaning Products

The easiest way to begin your transformation to a nontoxic home is by replacing cleaning products—ammonia, oven cleaners, furniture polish, scouring powder, disinfectant, glass cleaner—all the heavy-duty chemicals we use to maintain our homes. The replacement products are simple and inexpensive—in fact, most cleaning jobs can be done quite well using natural materials you probably already have in your kitchen. These substances are odorless or have subtle natural fragrances and work *every bit as well* as the chemicals you are accustomed to cleaning with. They also have the added benefit of being much less expensive than commercial cleaning preparations. You don't have to pay for advertising or packaging, or buy a different product for each cleaning need.

Why bother switching? Cleaning products are among the most hazardous products you will find in your home—so much so that they are regulated by the Consumer Product Safety Commission under the Federal Hazardous Substances Act.

This act decrees that products that can cause substantial personal injury or illness must carry warning labels. Such warnings are required on any product containing chemicals that are toxic, corrosive, flammable or combustible under pressure, radioactive, or that act as irritants or strong sensitizers.

Products that contain hazardous substances must have labels on which one of the following "signal words" is prominently displayed: DANGER (or POISON with skull and crossbones), WARNING, or CAUTION.

The Hazardous Substances Act focuses only on the *immediate* effects the product can have if not used according to instructions. Accidental ingestion, eye contact, prolonged skin contact, or breathing concentrated fumes of the product must all be guarded against. But a disturbing fact about cleaning products is that the chemicals they contain can also have devastating effects over time—and these warnings aren't mentioned on the label. That's how we are generally exposed to cleaning products: over time, daily, for years, even decades, and we get so accustomed to using these products that we forget they may be dangerous.

Most cleaning products can also be harmful during actual use, even when you follow the directions on the label to the letter. This isn't mentioned on the label, either, because it is not required by law. Toxic fumes from these products may produce slight reactions such as headaches, fatigue, burning eyes, runny nose, and skin rashes. However, even if you do not have an obvious response while using the product, after day-in, day-out exposure, year after year, your body may suddenly respond with cancer, heart disease, lung problems, or damage to the liver or immune system. Some substances in cleaning products may also cause birth defects and genetic changes.

The biggest problem in assessing the possible dangers of cleaning products is that the manufacturers are not required to list their *exact ingredients* on the label. You can't look at a label to be sure, for instance, that a certain furniture polish does not contain nitrobenzene, or that a mold and mildew cleaner is free from pentachlorophenol. The products on our shelves may contain many toxic chemicals, but we have no way of finding out what they are. Even the government and poison control centers cannot break the code of trade secrecy surrounding cleaning products. The best we can do is guess which chemicals *might* be in these products and evaluate their health effects.

Cleaning product manufacturers are also not required to inform us on the labels of the type of hazard associated with using the product or warn against product use by those who are at high risk because of specific medical conditions. For example, the labels on cleaning products in aerosol containers do not disclose that the aerosol mist can aggravate an existing lung condition, such as asthma. Asthma sufferers might have less trouble breathing if aerosol sprays and other volatile chemicals were not used in the home. The American Lung Association warns against the use of aerosol sprays, yet product labels do not reflect these specific health concerns.

If this is not enough, a study done by the New York Poison Control Center found that *85 percent of the product warning labels studied were inadequate.* Some labels list incorrect first aid information and others warn against dangers that don't even exist! Someone at my local poison control center told me a story of a child who had swallowed some silica gel, a relatively harmless crystalline substance. But because this product was incorrectly labeled as a poison *by the manufacturer,* the mother tried to induce vomiting by putting her finger down the child's throat (she should *never* have done this without calling a poison control center or doctor). The crystals came up and lacerated the child's esophagus, causing more harm than if they had just stayed in her stomach. Had the container not been mislabeled, the child would not have been harmed.

Nontoxic cleaning requires very few specialized ingredients. I do *all* my cleaning with a squirt bottle of 50/50 vinegar and water, liquid soap, and a can of nonchlorine scouring powder, available at most supermarkets and hardware stores. It couldn't be simpler.

You might need a few more items, though, to accomplish your specific cleaning needs. Here are some natural substances you might want to have on hand:

NONTOXIC CLEANING KIT

☐ **Baking Soda**
☐ **Salt**
☐ **Distilled White Vinegar**
☐ **Lemon Juice**
☐ **TSP** (Trisodium phosphate)—The makers of TSP create it by mixing soda ash (a naturally occurring mineral) and phosphoric acid (made by heating phosphate rock) to form disodium phosphate, and

then adding caustic soda (made by reacting naturally occurring minerals). TSP is moderately toxic by ingestion and is a minor skin irritant, but the big advantage of using it is that it doesn't produce toxic fumes. Keep TSP out of children's reach, and use gloves while cleaning with it, and it will not pose a hazard.

One environmental problem with using TSP is that it is a phosphate and contributes to the pollution of our waterways. However, given the choice, I prefer to use TSP occasionally (only when absolutely necessary) rather than a more toxic and harmful product.

TSP is widely available in hardware and paint stores.

☐ **Liquid Soap**—Look for a liquid *soap,* not detergent, preferably unscented. Your local natural food store will probably carry it. If not, you can easily order it by mail.

You can also make liquid soap from any bar soap. Mix 2 cups grated bar soap with 1 gallon water in a pot, and stir over low heat until the water boils and the soap dissolves. Lower the heat and simmer for 10 minutes. Cool and place in storage container.

☐ **Borax**—Available in any supermarket in the laundry products section.

☐ **Sodium hexametaphosphate**—This crystalline substance made from minerals has many unexpected uses, from laundry bleach to dishwasher detergent. It is not available in retail stores, but must be ordered from a chemical supply house. Look in the Yellow Pages of your telephone book for a local chemical supply house that will deliver sodium hexametaphosphate right to your door.

☐ **Nonchlorinated Scouring Powder**—Look in the cleaning section of your local supermarket for a scouring powder that says "No chlorine" on the label. You'll have to read the small print on the back of the can, but at least one national brand is so labeled.

Let's take a closer look at the dangers of some common cleaning products and how we can tackle these same cleaning chores with natural substances.

DRAIN CLEANERS 👄 ✍ 👤

From the Manufacturers' Warning Labels:

> **POISON: CALL POISON CENTER, EMER-
> GENCY ROOM, OR PHYSICIAN AT ONCE.
> Causes severe eye and skin damage; may cause
> blindness. Harmful or fatal if swallowed.**

The primary component of drain cleaners is lye, an extremely corrosive material that can eat right through skin. Even a drop spilled on your skin or a dry crystal that falls on wet skin can cause damage. When ingested, lye quickly burns through internal tissues, damaging the esophagus, stomach, and the entire intestinal tract. The internal damage may be irreparable for those who survive lye poisoning.

Lye itself poses no danger by inhalation, but in liquid drain cleaners lye is mixed with volatile liquid chemicals that can release harmful vapors.

WHAT YOU CAN DO

For all their dangers, lye-based drain cleaners are really not very effective. How many times have you tried to clear a drain with lye only to have the clog just sit there, leaving you with a sink full of corrosive lye-contaminated water and wondering what to do next? Why endanger the lives of your family for a product that doesn't even work?

My favorite drain-opener is the handy-dandy plunger—unlike lye-activated products, this old-fashioned standby works nearly every time. You can buy one that lasts for years for practically nothing at any hardware store. True, at times it takes more than a few plunges, but usually the clog will break down and float away eventually.

If this doesn't work, lye won't work either, so call a plumber. He or she will bring a long, flexible metal snake and push the clog away.

It's important, of course, to practice a little preventive plumbing. Use a drain strainer to trap food particles and hair, and remember not to pour grease down the drain (dump it in the garbage or into a "grease can" instead). These simple measures plus regular use of one of these nontoxic methods will keep drains open or clear sluggish drains before they become full-fledged clogs:

- Pour 1 handful baking soda and ½ cup white vinegar down the drainpipe and cover tightly for one minute. The chemical reaction between the two substances will form a fizzy pressure in the drain and dislodge any obstructive matter. Rinse with hot water.

- Pour ½ cup salt and ½ cup baking soda *or* 2 tablespoons trisodium phosphate (available in hardware stores) down the drain, followed by lots of hot water.

AMMONIA AND OTHER ALL-PURPOSE CLEANERS

From the Manufacturers' Warning Labels:

> **POISON: May cause burns. Call a physician. Keep out of reach of children.**
> **CAUTION: Harmful if swallowed. Irritant. Avoid contact with eyes and prolonged contact with skin. Do not swallow. Avoid inhalation of vapors. Use in a well-ventilated area.**

Ammonia is great for attacking household grease and grime, but it also attacks the skin—rashes, redness, and chemical burns during exposure are common. Because it is a very volatile chemical, its fumes are extremely irritating to eyes and lungs. Ammonia fumes can be especially harmful to anyone with respiratory problems. Children with colds or bronchitis and those with asthma will have greater difficulty breathing with ammonia in the air. Ammonia can also cause severe eye damage, if it is accidentally splashed in your eye, or if you absentmindedly rub your eyes with ammonia on your hands.

And if products come in **aerosol spray** cans the risk is increased, as the spray makes it much more likely that actual droplets of ammonia will end up on your skin, and in your eyes and lungs.

WHAT YOU CAN DO

You can create your own all-purpose cleaning solutions by mixing various natural substances together. You'll want to create your own proportions and combinations depending on the job. Start by mixing one teaspoon of any one or more of the following into 1 quart warm or hot water in a spray bottle or bucket: TSP (use for heavy duty cleaning), liquid soap, or borax. Add a squeeze of lemon juice or a splash of vinegar to cut grease. You'll soon find which combinations work best for your own cleaning needs.

If you want to buy a less-toxic commercial preparation, your natural food store should have a selection of soap-based all-purpose cleaners that do not contain ammonia.

OVEN CLEANER

From the Manufacturers' Warning Labels:

DANGER: CONTACT WILL CAUSE BURNS. Avoid contact with skin, eyes, mucous membranes and clothing. Do not take internally. Wear rubber gloves while using. Contains lye. If taken internally or sprayed in eyes call a physician. Keep out of reach of children. Irritant to mucous membranes. Avoid inhaling vapors. Contents under pressure. Recommended for use only on porcelain, enamel, iron, stainless steel, ceramic and glass surfaces. Do not get on exterior oven surfaces such as aluminum and chrome trim, baked enamel, copper tone, or painted areas, or on linoleum or plastics. For gas ovens, avoid spraying on pilot light. Keep spray off electrical connections such as door-operated light switch, heating element, etc.

Although oven cleaners contain several toxic ingredients, the greatest dangers come from **lye**, which can eat right through skin, and **ammonia**, which is extremely irritating to eyes and lungs.

Oven cleaners in **aerosol spray** containers are especially hazardous, because the spray sends tiny droplets of lye and ammonia into the air, where they can easily be inhaled, or land in your eyes or on your skin.

Be especially cautious about oven cleaners that advertise "no fumes." I have tried several brands and all still smelled very strongly of ammonia.

WHAT YOU CAN DO

If you are at all like me, you probably hate to clean your oven. It is possible to *never* have to clean your oven if you are very careful about not allowing things to spill. If your casserole looks like it might spill over during baking, put a cookie sheet or sheet of aluminum foil on the lower rack. On those rare occasions when something does end up on the bottom of the oven instead of on your plate, wipe it up as soon as the oven is cooled to prevent it from baking on even more.

Still, accidents do happen, and for those times, I'll give you a tip from a friend of mine who runs her own nontoxic cleaning service and refuses to use chemicals. Here's how Regina cleans ovens:

- Mix together in a spray bottle 2 tablespoons liquid soap (not detergent), 2 teaspoons borax, and warm water to fill the bottle. Make sure the salts are completely dissolved to avoid clogging the squirting mechanism. Spray it on, holding the bottle very close to the oven surface so the solution doesn't get into the air (and into your eyes and lungs). Even though these are natural ingredients, this solution is designed to cut heavy-duty oven grime, so wear gloves and glasses or goggles, if you have them. Leave the solution on for 20 minutes, then scrub with steel wool and a nonchlorine scouring powder. Rub impossible, baked-on, black spots with pumice, available in a stick at hardware stores.

If you have an extremely dirty oven layered with years of baked-on grease, you may have to use a chemical oven cleaner *once* to get it clean before you can begin your nontoxic maintenance. For this *one* application, choose a nonaerosol product and follow these precautions from the Consumer Product Safety Commission for safer use of oven cleaners:

- Read and follow the directions before each use.

- Wear protective gloves for your hands and goggles for your eyes.

- Open windows in the kitchen and be sure that children and other members of the family are out of the room.

- If you use an oven cleaner which requires boiling water, place the can in the oven before adding boiling water, so that you will not be overcome by ammonia fumes.

- If the fumes begin to irritate you, close the oven door, leave the room, and get fresh air.

FURNITURE AND FLOOR POLISH ☺ ◉ ✍ ↓

From the Manufacturers' Warning Labels:

> **DANGER: Harmful or fatal if swallowed. Keep out of reach of children.**

Regarding furniture and floor polishes, both accidental ingestion and inhalation of fumes during normal use (especially if an aerosol spray is used) are harmful. In addition, some of the more toxic ingredients can be easily absorbed through the skin. The primary danger is from exposure during use, but once it's on the furniture, polish can give off residual fumes.

Phenol, suspected of causing cancer in humans, is used in many furniture and floor polishes. Allowing phenol to touch your skin can cause it to to swell, peel, burn, or break out in hives and pimples. Even a small amount of phenol taken internally can cause circulatory collapse, convulsions, cold sweats, coma, and death. Keep this chemical away from your body!

Another chemical frequently used in furniture and floor polishes is **nitrobenzene**, which is extremely toxic and easily absorbed by the skin. An accidental spill on the skin can cause skin discoloration, shallow breathing, vomiting, and death. Repeated exposures can cause cancer, genetic changes, birth defects, and heart, liver, kidney, and central nervous system damage.

Because furniture and floor polishes are labeled "Harmful or *fatal* if swallowed," I would think these products could contain enough of these toxic chemicals to kill. Do you really want this stuff around your kids? Remember, they can't read warning labels and what will happen when one day they might want to imitate you (as kids love to do) and polish the furniture to surprise you?

Furniture and floor polishes might also contain harmful **acrylonitrile, ammonia, detergents, artificial fragrances, naphtha,** and **petroleum distillates,** and may be dispensed as **aerosol sprays.**

WHAT YOU CAN DO

It's easy to make your own furniture polish. The active ingredient in most polishes is plain mineral oil, which you can purchase at a drug store and apply sparingly with a soft cloth. Mineral oil is a petrochemical product that is relatively safe to use; its only real danger comes from repeated, regular ingestion. It's odorless and is absorbed right into the wood. Many polishes contain mineral oil—you'll just be using the same active ingredient without all the extra solvents and perfumes. If you like lemon-scented polish, you can add 1 teaspoon of lemon oil to 2 cups mineral oil.

The idea behind furniture polish is to get an oil absorbed into the wood, but it doesn't matter what kind of oil you use. It could be any oil, even one right in your kitchen cabinet. I just use plain mayonnaise. Just open the bottle, put a little on a soft cloth, and rub it in. Your furniture will smell like a sandwich for a few minutes, but the odor disappears quickly and the mayonnaise leaves a soft, nonsticky finish.

You can also mix your own "salad dressing" polishes:

- 1 teaspoon olive oil, mixed with the juice of one lemon, 1 teaspoon brandy or whiskey, and 1 teaspoon water. (Make fresh each time.)

- 3 parts olive oil mixed with 1 part white vinegar.

- 2 parts olive or vegetable oil mixed with 1 part lemon juice.

- For oak: Boil one quart beer with 1 tablespoon sugar and 2 tablespoons beeswax. Cool, wipe onto wood, and allow to dry. Polish when dry with a chamois cloth.

Don't worry about the odors any of these natural polishes might leave. The food smells quickly dissipate, and they don't become rancid.

SILVER POLISH AND OTHER METAL CLEANERS ☞ 👁 ✍ ⅃

From the Manufacturers' Warning Labels:

> **DANGER: Harmful or fatal if swallowed. Combustible. Irritating to eyes. Contains petroleum distillates and ammonia. If swallowed, do not induce vomiting. Call physician immediately. Keep out of reach of children. Keep away from heat and flame.**

The main chemicals of concern in metal cleaners are **ammonia**, which can burn the skin and produce irritating vapors, and unknown **petroleum distillates**. "Petroleum distillates" is not the name of a particular chemical, but rather of a whole group of chemicals of varying toxicity that are made by distilling petroleum. We have no way of knowing which petroleum distillate may be used in this product; all we are told is that it is fatal if swallowed. Since the manufacturer is required only to warn on the label of the dangers of accidental ingestion, we do not know if the vapors of this product might also be harmful, or if it can be absorbed through the skin.

WHAT YOU CAN DO

It is so easy to remove tarnish from silver without chemicals, I can't imagine why anyone would want to use smelly polish! Instead of rubbing for hours, you can *magnetize* tarnish away. The basic ingredients needed to perform this trick are aluminum (either in the form of a pot, pan, or aluminum foil) and some kind of salt (table salt, rock salt, or baking soda work fine). In the salty water, the aluminum will act as a magnet and attract the tarnish away from the silver. After submerging the pieces of silver for a few minutes in water containing both the aluminum and salt, you can literally wipe them dry and the tarnish will be gone (badly tarnished silver may need to go through the process several times). There are many ways this can be done. Here are a few methods you can try to see which is most convenient for you to work with:

- For very large items, such as trays or candelabra, run very hot water into your kitchen sink and add a sheet of aluminum foil and a handful of salt. Let sit for 2 or 3 minutes, then rinse and dry.

- For silverware, put a sheet of aluminum foil in the bottom of a pan, then add 2 or 3 inches of water, 1 teaspoon salt, and 1 teaspoon baking soda. Bring to a boil, and add silver pieces, making sure water totally covers the silver. Boil 2 or 3 minutes, then remove from pan, rinse, and dry.

- For small items, such as jewelry, fill a glass jar half full with thin strips of aluminum foil (if you have a manual pasta-making machine, making foil "spaghetti" works very well). Add 1 tablespoon salt and cold water to fill the jar. Keep covered. To use, simply drop small items into the jar for a few minutes, then remove, rinse, and dry.

To clean brass and copper, use lemon juice. Rub it on with a soft cloth, rinse with water, and dry. If this doesn't work, try applying one of the following mixtures using lemon juice:

- Make a paste of lemon juice and salt.
- Sprinkle a slice of lemon with baking soda.
- Make a paste of lemon juice and cream of tartar. Apply and leave on for 5 minutes before you rinse with warm water and dry.

If you don't have a lemon around the house, other kitchen staples clean brass and copper equally well: hot white vinegar mixed with salt, hot buttermilk, hot sour milk, tomato juice, and Worcestershire sauce.

Brass will look brighter and need polishing less often if rubbed with a bit of olive oil after each cleaning.

Clean chrome by wiping it with a soft cloth dipped in apple cider vinegar, or by rubbing with a lemon peel. Rinse with water and dry with a soft cloth.

There are a few nontoxic metal polishes on the market, but

they are generally very hard to locate. Check labels of polishes at your local supermarket and hardware store. If the polish is toxic, it is required by law to contain a warning label. Those without warnings are probably relatively safe.

MOLD AND MILDEW CLEANERS ⟨ 👁 ✋ 👃

From the Manufacturers' Warning Labels:

> **DANGER: Eye irritant. Keep out of reach of children. Use only in well ventilated area.**

Mold and mildew cleaners may contain **phenol, kerosene, or pentachlorophenol,** all of which may be harmful through skin absorption or inhalation, or fatal if swallowed. These products may also contain **formaldehyde,** suspected of causing cancer in humans and a strong irritant to eyes, throat, skin, and lungs.

The label warns that mold and mildew cleaners are a dangerous eye irritant, yet they are packaged in either pump or **aerosol spray** containers, sending the harmful mist into the air, where it can very easily end up in your eyes.

WHAT YOU CAN DO

You can make your own mold and mildew cleaner by mixing borax and water or vinegar and water in a spray bottle. Spray it on and the mold wipes right off. Borax also inhibits mold growth, so you might try washing down the walls in your bathroom with a borax solution and just leaving it on, or sprinkling borax in damp cabinets under the sink.

If you have a major mold problem, the best solution is to use heat. Put a portable electric heater in the room (any inexpensive quartz heater from the hardware store will work), turn it up to the highest setting, close the door, and let it bake all day or overnight. The mold will dry up into a powder that brushes right off. For concentrated areas, use a hand-held hair dryer to dry up the mold in just a few minutes.

Prevent mold growth by creating an environment in which it cannot live. Mold grows in dark, damp places, so keep rooms dry and light. During wet winters you may have to keep the heat on instead of wearing sweaters, if you live in a shaded area that gets little sun, or next to a creek, or any unusual location that is especially moist.

Allowing air to circulate will also help keep things dry. Hang clothes so there is space between them and, if you don't launder clothing damp with perspiration, at least allow it to dry before putting it back in the closet. Always hang wet towels after bathing to allow space for wetness to evaporate before throwing them in the hamper. If you have the space in your bathroom, you might want to try getting in the habit of hanging your wet towel on the towel rack after bathing, then replacing it right before your next bath or shower with a fresh one. Check the walls behind furniture to see if mold is growing there and rearrange furniture, if necessary, to allow plenty of air flow. Small fans can increase air flow, if needed.

GERM-KILLING DISINFECTANTS

From the Manufacturers' Warning Labels:

> CAUTION: Keep out of reach of children. Keep away from heat, sparks, and open flame. Keep out of eyes. Avoid contact with food.

Disinfectants contain a number of volatile chemicals that are dangerous to inhale. The ingredient most frequently found in disinfectants is **cresol**, a chemical easily absorbed through the skin and through the mucous membranes of the respiratory tract. Cresol can cause damage to the liver, kidneys, lungs, pancreas, and spleen, and can also affect the central nervous system, resulting in such common symptoms as depression, irritability, and hyperactivity. Experts warn that there is danger of poisoning by both ingestion and inhalation. Disinfectants pose a hazard even in their closed containers, since vapors can leak out and build up to high levels in closed rooms.

Disinfectants may also contain other toxic germ killers: **phenol, ethanol, formaldehyde, ammonia,** and **chlorine.**

Ironically, most of us use a disinfectant when there is an illness in the family, just when the sick person is most vulnerable to toxic effects. Some people are so concerned about cleanliness that they overdisinfect, and are constantly surrounded by that "fresh, clean smell"—actually dangerous fumes.

WHAT YOU CAN DO

You may not even need to use a disinfectant. Like the television commercials say, disinfectants will "kill germs on contact," but they will not kill all the germs, only some of them. Germs are very friendly with one another and will quickly multiply soon after disinfecting.

If you want to eliminate disease-causing bacteria and viruses, you must sterilize items by immersing them in boiling water. To reduce germs on large surfaces (such as walls or floors), try using a solution of ½ cup borax (available in the cleaning products department of your supermarket) dissolved in 1 gallon hot water. One hospital experimented with this mixture for one year, and the monitoring bacteriologist reported that it satisfied all the hospital germicidal requirements.

Regular cleaning with plain soap and water will keep germs under control. Even just a rinse of hot water will kill some bacteria.

If, for some reason, you need a liquid disinfectant, the safest I know of is an aqueous solution of benzalkonium chloride 1:750 (available at drug stores under the brand name Zephirin). It is usually used for medical purposes, but it can be substituted for any liquid disinfectant and produces far fewer fumes.

RUG, CARPET, AND UPHOLSTERY SHAMPOO

From the Manufacturers' Warning Labels:

> **CAUTION: Do not take internally. In case of eye contact, flush thoroughly with water. Keep out of reach of children.**

The active ingredient likely to be found in most rug, carpet, and upholstery shampoos is **perchloroethylene,** a solvent commonly used as a spot remover. It is a known human carcinogen and its immediate effects can be light-headedness, dizziness, sleepiness, nausea, tremors, loss of appetite, and feelings of disorientation. Long-term exposure may result in damage to the liver or central nervous system.

Rug, carpet, and upholstery cleaners may also contain **naphthalene,** which, according to the chemical dictionary, is "toxic by inhalation." Headaches, confusion, nausea and vomiting, excessive sweating, and urinary irritation can all result from exposure. Naphthalene is also suspected of causing cancer in humans.

Ethanol, ammonia, and **detergents** are other common ingredients in these products.

Rug, carpet, and upholstery shampoos frequently leave residues after use. While I have heard no reports of the possible health effects of these residues, I don't think I would want my baby to be crawling around on a carpet cleaned with such dangerous substances.

WHAT YOU CAN DO

The safest carpet cleaners on the market are those with a baking soda base, but these are really only scented baking soda in fancy packages. Regular unscented baking soda (the kind you use for cooking) works just as well, is less expensive, and doesn't contribute to indoor air pollution with artificial fragrance.

To deodorize your carpets with baking soda, sprinkle it liberally over the entire carpet (make sure the carpet is dry first). By "liberally," I mean it should look as if it had snowed on the carpet. You will need several pounds for a $9' \times 12'$ room. Wait 15 minutes or longer, then vacuum. You can even leave it on overnight. I tried this once with an old wool oriental area rug that had been sitting in my great-aunt's attic. It had that usual musty, moldy smell, and after two applications of the baking soda, the odor was entirely gone.

To brighten the color of your carpet, vacuum first to remove dust, then apply a mixture of 1 quart white vinegar and 3 quarts boiling water with a wet rag. It will smell like vinegar when it's wet, but will dry odor-free and is a lot safer than using chemicals. Be careful to just dampen the nap of the rug, and not to saturate the backing. Allow the rug to dry thoroughly, then rub the surface with breadcrumbs and vacuum.

Try to wipe up spills immediately, before they become stains. Plain water will generally work, but if you need a stronger spot remover, try a solution of ¼ cup borax dissolved in 2 cups cold water, or undiluted vinegar or lemon juice.

DISHWASHER DETERGENT

From the Manufacturers' Warning Labels:

> CAUTION: Injurious to eyes. Harmful if swallowed. Avoid contact with eyes, mucous membranes, and prolonged skin contact. Keep out of reach of children.

As with many other cleaning products, the primary danger of dishwasher detergent is accidental ingestion and, to a lesser degree, exposure during use. Most dishwasher detergents contain chlorine in a dry form that is activated when it comes into contact with water in the dishwasher, releasing chlorine fumes into the air that leak out of the dishwasher and into the kitchen. While low levels of chlorine are considered safe for most, many people report such symptoms as headache, fatigue, burning eyes, and difficulty breathing when exposed to even the small amount of chlorine released during normal dishwashing.

Also, a very thin film of detergent can remain on the washed dishes. Run a load of dishes through your dishwasher and compare them to dishes washed by hand. You may be able to see, smell, feel, or even taste the difference. If there is a residue, you are ingesting dishwasher detergent in small amounts every time you eat or drink. The long-term health effects are unknown.

HOW TO PROTECT YOURSELF

Using sodium hexametaphosphate in your dishwasher works well. I know this sounds like an intimidating chemical, but it's just a mild mineral powder that dissolves grease and doesn't leave water spots on the glasses. Its one disadvantage is that it won't remove dried-on food, but for regular washing, it works just as well as any commercial chlorinated brand. You can order sodium hexametaphosphate by mail from most chemical supply houses (look in the yellow pages of your telephone book for one near you).

You could also try to find one of the few brands of dishwasher detergent that does not give off strong chlorine fumes. These should still be stored safely and used with caution as they contain other harmful ingredients.

NOTE: Never use dishwashing liquid as a substitute in a dishwasher, as the bubbles can clog the drain and inhibit the action of the water spray.

CHLORINATED SCOURING POWDER

Almost all scouring powders contain **chlorine** bleach, which acts as a whitener and stain remover. The powdered chlorine produces chlorine fumes when it comes in contact with water, which can be highly irritating to eyes, nose, throat, and lungs. Some people also react with headaches, fatigue, and difficulty breathing.

One warning that should be on the scouring powder can (but isn't) is the same one you'll find on a chlorine bleach bottle: *Never mix chlorine with ammonia; the resulting chloramine fumes can be deadly.* Ammonia is hidden in many cleaning products, so you may not always know that you're mixing them. You could, for instance, use an all-purpose cleaner with ammonia in your toilet bowl, then sprinkle in some scouring powder. The combination is of course less of a hazard than if you poured chlorine bleach into the toilet bowl with the ammonia; but even the level of chloramine fumes produced by ammonia and scouring powder could be dangerous, depending on your (or your child's) individual ability to tolerate these fumes.

Detergents and **talc** are also used in scouring powders. The danger of talc is that it may be contaminated with carcinogenic **asbestos,** and when you sprinkle scouring powder in your sink, a small amount always goes into the air, and possibly into your lungs. There is *no* safe level for asbestos exposure.

WHAT YOU CAN DO

There is at least one nationally available brand of scouring powder on the market that does not contain chlorine or talc. It lists its ingredients on the back label, so you should have no trouble finding it.

Dry baking soda, borax, or table salt sprinkled on a wet sponge also act as effective abrasives for scouring.

GLASS CLEANERS

Most glass cleaners are nothing more than **ammonia** mixed with water and a little **blue dye.** It is interesting that ammonia bottles are labeled "POISON," yet glass cleaners have no warning labels at all! Glass cleaners containing ammonia can release highly irritating fumes, and can cause eye damage if accidentally sprayed into the eye.

Glass cleaners in **aerosol spray** containers are even more dangerous, because the aerosol spray distributes tiny droplets of ammonia into the air, where it can be easily inhaled or float into your eyes.

WHAT YOU CAN DO

I am only going to recommend one glass cleaner because it works so well, I don't know why you would want to try anything else: half water and half vinegar in a pump spray bottle (or you can put it on with a sponge from a bucket). I wipe it off with an old cotton terrycloth towel, but any rag will do.

You might run into one problem with this and it will only happen the first time you use it. Some chemical glass cleaners contain a wax that can build up on the glass and cause it to streak terribly when you use vinegar and water. If you have been using this type of glass cleaner you might have years of build-up that will have to be removed before you can use the natural cleaner. Use a little rubbing alcohol to get it off, and then you can clean your windows chemical-free!

AIR FRESHENERS ⊖ ✑ ℒ

Most air fresheners don't "freshen" the air at all—they cover up the offensive odor with a more pleasant one, or interfere with your ability to smell by releasing a nerve-deadening agent, or coating your nasal passages with an undetectable oil film.

While little (if any) scientific research has been done on the health effects of air fresheners, they are made from a number of chemicals known to be toxic in amounts larger than is found in an air freshener: **naphthalene, phenol, cresol, ethanol, xylene,** and **formaldehyde.** Many people with allergies and respiratory problems are bothered by the smell of air fresheners; the possible effects of long-term, low-level exposure to these chemicals is unknown.

WHAT YOU CAN DO

Air freshener is one of those products that is highly advertised but probably completely unnecessary.

You can rid your home of undesirable odors simply by opening a window or turning on a fan. This will also help reduce any toxic fumes that are building up indoors.

Track down odors in your home and find out what's causing them. For mold smells, keep the area dry and light and, if necessary, use a small bag of silica gel (you can buy it at a camera store) to absorb excess moisture from the air. Silica gel changes color when it is saturated; you can reactivate it by drying it out in the oven.

To reduce food smells, empty the garbage frequently and clean the can when needed. Sprinkle ½ cup borax (in the cleaning products department at your supermarket) in the bottom of the garbage can to inhibit the growth of odor-producing mold and bacteria.

If you use air fresheners to scent the air, try natural fragrances, which are even more pleasant. Why not use the real thing—fresh flowers or pine boughs—to provide a delicate scent? Or you can make sachets of fragrant herbs and flowers. Go to the herb section of your natural food store and sniff the different jars. I like the smell of ground cloves in my kitchen in the wintertime, and peppermint in the bathroom—just put them in a small basket or jar, or in a small sachet bag.

You can make your whole house smell wonderful, and it costs a lot less too.

DISHWASHING LIQUID

Dishwashing liquid is just a liquid **detergent** with some **dye** and **artificial fragrance.** Sounds safe enough and there are no warning labels, but it is still a detergent and according to the Center for Science in the Public Interest, *detergents are responsible for more household poisonings than any other household product.* Their fruity smells are particularly attractive to children. Here's another case of a product we use daily and usually consider totally benign, which actually belongs up high and locked away in the "childproof cabinet."

I am also concerned about the coloring used in liquid detergents. I have been unable to find any information that indicates these dyes are regulated for safety in any way. Dishwashing detergent is not a food, drug, or cosmetic, so these colors are not regulated by the FDA. The Consumer Product Safety Commission regulates cleaning products, but they are concerned only with the "hazardous ingredients." Some artificial dyes used in food products are allowed even though they are known to cause cancer; it makes me wonder about the safety of soaking your hands in these colors three times a day.

WHAT YOU CAN DO

Use a plain liquid soap, available at most natural food stores. Or you can rub your sponge with bar soap, although I have found with this method you have to be very careful not to drop the dishes—they get very slippery!

If you like lemon-scented dishwashing liquid, use a few slices of fresh lemon in the dishwater. This will also act to help cut grease.

If you live in a hard-water area and you end up with water spots on your glasses, add a few teaspoons of sodium hexametaphosphate (order from a chemical supply house) to your dishwater and use only about half the amount of soap you would normally use. Your dishes will dry spotless!

CHAPTER 3

Household Pesticides 👄 👁 ✋ 🦵

The best way to control pests is by natural means. Natural pest control methods are even more effective than chemical pesticides because many pests are beginning to develop immunities to the chemicals. Some of our worst pests are now resistant to practically all our chemical weapons. So we come back to Mother Nature and common sense as our best defense.

Household pesticides contain chemicals specially formulated to kill. Fly sprays, flea bombs, ant and roach killers, and rat and mouse poisons are used in or around 91 percent of all American households. Of approximately 1,400 different active pesticide ingredients used in more than 35,000 formulations, more than 100 are known to cause cancer, birth defects, and mutagenic changes, although authorities believe there could be more, since most pesticides have not been adequately tested for these effects. Poison control centers report that many people die from exposure to pesticides (both from accidental ingestion of liquids in storage and inhalation of sprays during normal use)—and more than half of the fatalities are children.

Not only is a pesticide hazardous during the actual application, but residues can stay in an area for a very long time. Pesticides can remain active in the air for days or weeks; some can even last up to twenty years! When several types of pesticides are used, residues can intermingle, and when they combine, these become even more toxic. It's frightening to consider what toxic compound could be created from the pesticides in your food mixed with the pesticide residues in the air, mixed with pesticides in your tap water, mixed with your pet's flea collar, etc. Fortunately, we can choose not to use these products.

Pesticides are stored in the fatty tissue in the body, and can accumulate over time to dangerously high levels. When you exercise and "burn fat" or go on a weight reduction diet, pesticides are released into the bloodstream.

Immediate health effects from inhaling some common household pesticides during use include nausea, cough, breathing difficulties, depression, eye irritation, dizziness, weakness, blurred vision, muscle twitching, and convulsions. Long-term exposure from repeated use and lingering residues can also damage the liver, kidneys, and lungs, and can cause paralysis, sterility, suppression of immune function, brain hemorrhages, decreased fertility and sexual function, heart problems, and coma.

The Environmental Protection Agency (EPA) regulates all pesticides under the Federal Insecticide, Fungicide, and Rodenticide Act (FIFRA), which controls the manufacture, use, and disposal of poisons used in agriculture, forestry, household, and other activities. This act requires that all pesticides be registered, it classifies pesticides for household or restricted use, it gives the EPA the power to ban harmful pesticides, and it requires informative and accurate labeling.

As well-regulated as pesticides are, product labels give us very little information. All that must be listed are the *active* ingredients that actually function as poisons. Generally, their names are long and intimidating (even to me!) and give us no information as to their safety.

The simplest way to assess the danger of a pesticide is to look at the signal word warnings on the label:

Those marked with **DANGER—POISON** with skull and crossbones could kill an adult if only a tiny pinch is ingested.

Those marked with **WARNING** could kill an adult if about a teaspoon is ingested.

Those marked with **CAUTION** will not kill until an amount from 1 ounce to 1 pint is ingested.

The EPA was unable to furnish me with a list of pesticides used in household products, or examples of pesticides that fit into their signal word categories. But here is a list of pesticides' relative toxicities from the book *Bugbusters: Getting Rid of Household Pests Without Dangerous Chemicals* by Bernice Lifton (New York: McGraw-Hill Paperbacks, 1985), based on a list compiled by the University of California at Berkeley:

MOST DANGEROUS aldicarb, carbofuran, demeton, disulfoton, fensulfothion, fonophos, methamidophos, mevinphos, parathion, phorate, schradan, TEPP.

DANGEROUS aldrin, bufencarb, carbophenothion, chlorpyrifos, DDVP, dichlorvos, dicrotopohos, dieldrin, dinitrocresol, dioxation, DNOC, endrin, EPN, methylparathion, mexacarbate, monocrotophos, nicotine, paraquat, pentachlorophenol, phosalone, phosphamidon, propoxur.

LESS DANGEROUS akton, azinphosmethyl, binapacryl, BHC, chlordane, chlordimeform, caumaphos, crotoxyphos, crufomate, diazinon, dicapthon, dichloroethyl ether, dimethoate, dinobuton, endosulfan, ethion, fenthion, heptachlor, lead arsenate, lindane, metam sodium, naled, oxydemeton methyl, phosmet, toxaphene, trichlorfon.

LEAST DANGEROUS aminozide, captan, carbaryl, chinomethionat, 2, 4-D, daminozide, DDT, dicofol, diquat, malathion, maneb, naphthaleneacetic acid, oxythioquinox, Perthane, propargite, quinomethionate, ronnel, rotenone, TDE, temephos, tetrachlorvinphos, tetradifon.

I found a general-purpose household pesticide spray on a local supermarket shelf containing DDVP as its active ingredient. It was labeled "CAUTION" by the EPA, and DDVP is listed as "DANGEROUS" on the University of California list. Picking your way through this semantic jungle and trying to figure out how dangerous any product is are ultimately not worth the effort, because available nonchemical alternatives are *every bit as, if not more, effective!*

The key to success in any natural pest control program is to prevent pests from entering your home and to make it impossible for them to live there. If bugs are driving you nuts, it's probably because you are putting out the welcome mat by providing them with perfect living conditions. If you don't take these two steps, the best you can do is get rid of the pests temporarily, for they will return again and again. But once you have pest-proofed your home, you will be permanently protected from almost all unwelcome visitors.

The first thing to do is figure out how pests are getting into your home, and do something to keep them out. If you have ants, follow the line of ants around to see where they're coming in, then seal up the hole with white glue (it's less toxic than caulk). They're sure to find another entry point, but after a few days of keeping a sharp lookout and wielding a swift glue bottle, you'll fill all your holes and eliminate the ant problem for good. Also, you might need screens on your windows to block flying insects. And maybe you need to look for and fill holes and cracks in your building structure that mice can crawl through.

Second, make your home an unpleasant place for pests to be. Here's how:

- Take away their food supply by keeping living areas clean. Be especially careful to sweep up food crumbs (don't leave your cracker crumbs in bed!), wipe up spills when they happen, wash dishes immediately, and deposit leftover food in its proper place. Store food in tightly closed, impenetrable metal or glass containers. Empty garbage cans frequently and, if necessary, accumulate garbage in a plastic bag with a twist tie to eliminate the enticing odors of decaying food. Dispose of disposable diapers using good hygiene, as they, like food waste, attract pests.

- Dry up their water supply. Repair leaky faucets, pipes, and clogged drains. Insects have to drink somewhere—don't let poor plumbing turn your home into the "neighborhood bug bar."

- Get rid of any clutter that they can hide in. Clean out your attic, basement, and closets. Remove piles of old clothing, newspapers and magazines, and boxes. Check out-of-the-way places especially well (under stairwells, for instance) where unused items often get tossed.

You can eliminate the pests already living in your home with homemade potions made from natural ingredients you probably already have right in your kitchen. Or you can use mechanical methods to control them such as zapping them with extreme temperatures, using fragrances that are offending (to them, not to you!), or trapping them. These methods are more effective than chemicals, easy to do, and safe for children and pets.

RAT AND MOUSE KILLERS 👄 👁 ✋ 1

From the Manufacturers' Warning Labels:

> **DANGER—POISON: Keep out of reach of children. This bottle contains a deadly poison—arsenic. Poisonous if swallowed. Do not get in eyes, or on skin, or clothing. Wash thoroughly after handling.**

Rat and mouse killers are the most harmful pesticides available for home use. They can contain **arsenic, strychnine,** or **phosphorus,** all of which kill quickly if ingested.

WHAT YOU CAN DO

Would you think the solution to a rat and mouse problem would be to get a cat? Cats will in fact help reduce the number of rats and mice in the area, but only by about 20 percent.

The old-fashioned mouse trap is the best way to catch rats and mice. The preferred bait is not cheese, but peanut butter sprinkled with a bit of cornmeal or oatmeal. Use plenty of traps; for an average home, twelve is probably plenty. It's better to use more traps for a short period of time than to use one trap and expect each rodent to eventually pass by. Rats and mice are very clever and have a way of avoiding traps, especially if they're always in the same place. So keep your bait fresh, use enough traps, and change their location every few days for maximum effectiveness. You might try placing your traps in open paper bags. If your rodents can still find the traps, it will make your disposal job much easier.

Another way to kill mice is by sprinkling an attractive poison around the area. Instead of using a chemical, make a mixture of 1 part plaster of Paris and 1 part flour with a little sugar and cocoa powder added to make it taste good. They eat it, and they're gone.

If you have a major problem with rodents, call a professional pest control company for detailed advice on how to "rodent-proof" your home. They can give you good advice on how you can block pests from coming into your home, but don't let them sell you on the whole poison routine. Just say you want to try whatever can be done without chemicals first.

INSECTICIDES 👄 👁 ✋ 👃

From the Manufacturers' Warning Labels:

> **CAUTION: Keep out of reach of children. Use only when area to be treated is vacated by humans and pets. Not to be taken internally by humans or animals. Hazardous if swallowed or absorbed through skin. Do not get on skin, in eyes, or on clothing. Avoid breathing of vapors or spray mist. Do not smoke while using. Should not be used in homes of the seriously ill or those on medication. Should not be used in homes of pollen-sensitive people or asthmatics. Do not use in any rooms where infants, the sick, or aged are or will be present for any extended period of confinement. Do not use in kitchen areas or areas where food is prepared or served. Do not apply directly to food. In the home, all food processing surfaces and utensils should be covered during treatment or thoroughly washed before use. Remove pets and cover fish aquariums and delicate plants before spraying.**

According to the EPA, no insecticides that are sold for general home use have warning labels stronger than "CAUTION." Any products that pose greater danger are available only for application by licensed pest-control operators.

Most insecticides kill all types of flying and crawling insects: ants, fleas, cockroaches, mosquitoes, flies, and silverfish. One of the most commonly used pesticides in home sprays is pyrethrin, which is just the crushed, dried flowers of *Chrysanthemum cinerariifolium.* It's harmless for humans or pets to swallow, but kills bugs on contact. The problem with commercial pyrethrin formulations is not the pyrethrin, but the fact that it is mixed with other pesticides, frequently dispensed in aerosol sprays, and diluted with "inert ingredients" of unknown type of toxicity.

WHAT YOU CAN DO

You can order pure pyrethrin powder by mail, or if you have a patch of garden, you can grow your own *Chrysanthemum cinerariifolium.* When two or three outer rows of petals have opened, gather blossoms and dry either in the sun or in an oven set at the lowest temperature. Grind dry flowers into a powder using a coffee mill, blender, or mortar and pestle. Place 10 grams of powder in a light-proof bottle and add 4 ounces alcohol. Shake the mixture occasionally, and let it stand for 24 hours at room temperature. Pour the finished mixture through coffee-maker filters and use as is, spraying through a small spray bottle, for roaches, flies, fleas, beetles, lice, and other insects.

I know making your own pyrethrin spray is a bit complex and requires some advance preparation, so here are some tips on how you can handle your pest problems today.

Ants

Wipe ants up with a wet sponge when you see them. They rely on each other for direction, so without their leader, the others will get lost.

To keep ants out, sprinkle powdered red chili pepper, paprika, dried peppermint, or borax where ants are coming in. Or use the botanical approach: Plant mint around the outside of the house, and they won't even think of coming inside, because—silly things!—they don't like the smell.

Cockroaches and Silverfish

Cockroaches are especially fond of vegetables, meat, starches, grease, sweets, paper, soap, cardboard, bookbindings, ink, shoe polish, and dirty clothes. And they're not fussy about their places of residence either: They might be living in your telephone, electric clock, radio, or even your refrigerator. So how do you get rid of them?

If you have a whole houseful, try baking them out. According to the British Museum of Natural History, raising the temperature to 130° F will successfully stop a roach infestation. Close all the windows, turn up the heat full blast, and leave the house for a few hours (you will end up with crispy roaches, but they won't be jumping around, or multiplying). I've never tried this, but I am told your house will not catch fire. At the other extreme, roaches cannot live at temperatures below 23° F. It might be harder to take advantage of this fact, unless you live in a very cold climate and try it in the dead of winter.

For limited infestations, use cockroach traps—either buy the commercial "motels" with the adhesive strip, or make your own (see instructions below). Use several in each room where you've seen roaches. When the traps start working, leave a few cockroaches inside to attract others. Clean traps by immersing them completely in a bucket of hot, sudsy water; and then make sure to wash your hands thoroughly.

Now, here's how to make your own wicked but nontoxic roach traps:

- **Trap #1:** Rub grease on the inside of the neck of a quart Mason jar. Set the jar upright, and put a piece of banana inside for bait. Place a tongue depressor against the side of the jar and the cockroaches will "walk the plank" to their deaths.

- **Trap #2:** Wrap masking tape around the outside of an empty jam jar. Fill the jar half-full with a mixture of beer, a few slices of banana, and a few drops of anise extract. Or use boiled raisins, pet kibble, or pieces of apple, potato, or banana peel as bait. Finally, smear a band of petroleum jelly around the inside rim so the cockroaches can't climb out.

- **Trap #3:** Soak a rag in beer and place it in a shallow dish overnight in an infested area. In the morning, you can easily dispose of the drunken roaches.

You could also put cucumber rinds or bay leaves in the infested area, or spread around any of the following mixtures. Repeat weekly for several weeks to kill newly hatched roaches.

- Equal parts flour or powdered oatmeal mixed with plaster of paris.
- Equal parts baking soda and powdered sugar.
- 2 tablespoons trisodium phosphate (TSP), ¾ cup borax, ½ cup granulated sugar, mixed with 1 cup flour.
- 2 tablespoons flour, 1 tablespoon cocoa powder, with 4 tablespoons borax.

If none of the above methods works (and you probably won't need to do this), as a last resort use boric acid, available in hardware and building supply stores. Make sure to get *technical* boric acid and not *medicinal* boric acid. Medicinal boric acid is a white powder that can easily be confused with sugar or salt; a tablespoon accidentally eaten can kill a small child. Technical boric acid, used in industrial manufacturing, is just as dangerous but is tinted blue for easy identification. It also has the added benefit of being treated electrostatically, so it will cling to the cockroach—when other cockroaches rub against the roach, they'll pick it up. There are several brands of commercial roach killers made of 100 percent boric acid, but buying plain boric acid is cheaper. Sprinkle it around in out-of-the way corners and it will continue to kill roaches as long as it is there. Boric acid is preferable to chemical sprays because it is a nonvolatile powder, but it is still a poison, so don't use it around food or in cabinets where food is stored, or where children or pets may come in contact with it.

Everything that works for cockroaches also works for silverfish.

Fleas

Whatever you do, do not use chemical sprays, powders, or collars on your animals. Cats and dogs (and their owners) are frequently poisoned by these pesticides. In fact, most pet poisonings are the result of the organophosphate and carbamate compounds used to control fleas.

Antiflea sprays contain toxic solvents that cause the insecticide to penetrate the skin—both your pet's and yours when you touch the animal. Powdered products can easily end up in your lungs, or in

47

your cat's stomach from licking its fur. Flea collars create a toxic cloud around your animal 24 hours a day, which you and your children breathe every time you go near your pet.

If you already have an infestation, vacuum frequently, daily if necessary, disposing of the vacuum bag after each cleaning. Use the crevice attachment to get in all the corners of both floors and upholstery. To speed up the process, have your carpets and upholstery steam-cleaned. Most pest control places will say this won't work and recommend fumigation, which involves closing up your home and spraying the entire interior with highly toxic chemicals. Before you agree to this expensive procedure, remember babies, pets, and children will be crawling around on a chemically-treated rug and their defense systems do not easily assimilate these toxic substances.

The easiest way to kill fleas is to heat your home to 122° F for several hours. Just take your pets and plants outside, close the windows and turn up the heat full blast. Take your family on an outing for the afternoon, then come back and vacuum everything.

As a preventive measure, you can actually make your pet flea-resistant by feeding it brewer's yeast. The yeast gives an odor to the animal's skin that the fleas find unpleasant. Use 25 milligrams per 10 pounds of the animal's body weight, beginning in the spring and continuing through the warm season. To prevent intestinal gas, feed the yeast to your pet in small amounts with moist food. If your pet is allergic to yeast, try rice-based B-complex vitamins available in natural food stores.

In addition, natural food stores stock many herbal repellants. Perhaps the easiest approach of all is to rub ground cloves, eucalyptus oil, or strong wormwood tea (wormwood leaves can be purchased from an herbalist or natural food store) directly into the animal's fur. This seems to work better than the herbal collars available at health food stores, because the repellant is more evenly distributed around the body.

Another very effective repellant is citrus oil. Studies have shown the citrus oil discourages insects of all types quite effectively. Here's how to turn plain lemons into a powerful flea-repellant: Place four cut lemons in a saucepan, cover with water, and bring to a boil. Simmer for 45 minutes. Cool and strain the liquid, and store in a glass container. Apply the liquid liberally to the animal's fur while brushing its coat so the citrus oil penetrates all the way down to the skin. Dry with towels and brush again.

Remember to bathe your pet frequently and use a flea comb (available at a pet store) at bath time. A comb, in addition to natural repellants, should keep your pet (and your home) permanently flea-free!

Flies

During the day, you can encourage flies to leave by darkening the room and opening the door to the outside light. Because flies are attracted to light, they will quickly fly outdoors.

Repel flies by hanging clusters of cloves around the room, or by scratching the rind of an orange or lemon to allow the unappealing (to them) citrus oil to escape. You will hardly notice the pleasant aroma.

If a fly happens to come in anyway, keep a flyswatter on hand, or make your own flypaper. Simply boil equal parts sugar, corn syrup, and water together and place it on brown paper. The flies can't resist this sticky treat. If this seems like too much trouble, you can buy adhesive fly strips at most hardware stores.

Food Storage Pests (Beetles, Weevils, Moths, and Mites)

Food storage pests enter the home through contaminated groceries. Inspect all packaged foods at the store before putting them in your cart. Look for loose package flaps and tiny holes in the packaging. Be particularly watchful with any grain products (including whole grains, flour, cereals, pasta, baking mixes), dried fruits, beans, powdered milk, and pet foods. Buy only small amounts of these products, which can be eaten in a short period of time.

Store foods in tightly closed containers in a cool, dry cabinet, or in the refrigerator. Packets of silica gel (available at building supply stores) will absorb moisture and help keep cabinets dry. Check food supplies once a week in summer and once a month in winter, and throw away (in the outside garbage) any contaminated foods before the infestation spreads to other containers. If you buy foods in bulk, you might think about getting an extra refrigerator to put in the garage for storing extra supplies. Not only will the foods be protected from insects, but they will stay fresher, too.

Putting a bay leaf in each container will help to repel stray

49

pests, but will not eradicate an infestation. If you want to try this, put the bay leaves in small cotton bags to prevent them from crumbling into the food. You wouldn't want to bake a cake with bay-flavored flour.

Mosquitoes

The main strategy for dealing with mosquitoes is to keep them from biting you and from buzzing in your ear when you are trying to sleep. So if you don't care if they're still flying around as long as they're not bothering you, dip a cotton ball in some vinegar and rub it on your exposed skin. The vinegar evaporates immediately and, no, you won't smell like a pickle. Do not use commercial insect repellants containing "Deet"—common name for diethyl toluamide. Not only is it an irritant to sensitive skin and respiratory tract tissues, but it can eat right through plastics and dissolve paint. Hardly something you would want to rub repeatedly into your skin.

Two other natural repellants are oil of citronella and oil of pennyroyal mint, both available at most natural food stores. These are very strong, however, and can result in a rash when applied directly to the skin and bad irritation to eyes when rubbed in accidentally. Dilute these oils with vodka or vegetable oil (a few drops to one ounce of either) and then apply them, like perfume, at strategic points.

Mosquitoes also hate the smell of garlic. The recommended method for using this repellant is to eat lots of garlicky food. The trick here is to get all your friends to eat a lot of garlic so none of you will notice each other's bad breath—or at least none of you will care—and you'll all be protected from mosquito bites.

If you don't even want the mosquitoes in the same room with you, wait until they light upon a wall and then suction them up with the long attachment on your vacuum cleaner.

Basil planted outside your window will prevent mosquitoes from coming in, or you can keep a few basil plants inside. Not only will you keep mosquitoes away, but you'll have an endless fresh supply of this wonderful herb. Once you've tasted fresh basil, you'll never go back to using the dried leaves in the jar.

For a severe mosquito problem, pretend you live in an old movie and get some romantic-looking mosquito netting for your bed.

LICE SHAMPOO ☙ ◉ ☞ ℩

Every parent with a school-age child might someday have to deal with a child picking up head lice at school. Contrary to what we would like to believe, head lice is a condition that cuts across all social and cultural boundaries—it can show up on anyone.

At one time the most common treatment for head lice was a shampoo that contained **lindane,** a very toxic chemical easily absorbed through the skin. One child's death was reported to have been the result of lindane poisoning after treatment for head lice, and lindane is now known to cause convulsions, seizures, and cancer in laboratory animals. When I checked my local drugstore shelves none of the half dozen shampoos available contained lindane. Nevertheless, keep an eye out for this very dangerous chemical.

The popular active ingredient nowadays is pyrethrin. As with general-purpose pesticide sprays, the toxicity problem lies not with the pyrethrin, but with the added petroleum distillates. Because the petroleum distillates frequently are not identified, I would hesitate to put chemicals of unknown toxicity on my child's scalp, especially since the scalp area is very porous, and chemicals put on the hair are easily absorbed into the bloodstream.

WHAT YOU CAN DO

A very effective, and much less toxic, alternative is to use a combination of shampooing, soaking, and combing with a nit-removing comb especially designed to eliminate lice (available at your local pharmacy). Here is the procedure:

1. Wet hair thoroughly with warm water and apply a coconut-based shampoo. Coconut oil contains dodecyl alcohol, which is deadly to adult lice (any shampoo that lists sodium lauryl sulfate as an ingredient contains coconut oil. Or you can use a bar soap that has a coconut-oil base). Work the soap or shampoo into a thick lather, covering the entire head and all the hair.

2. Rinse with warm water and repeat the lathering process, this time leaving suds on hair. Tie a towel around the lathered hair and leave it on for 30 minutes.

3. Remove the towel and comb the soapy hair with a regular comb to remove tangles; then use the nit-removing comb on 1-inch sections of hair (following the instructions that come with the comb). If the hair dries during combing, dampen it with water. Depending on the length of hair, the combing can take two hours or more. Very curly or woolly hair can take even longer.

4. After removing lice, wash the hair a third time, rinse, and dry. Inspect the hair when dry for any lice you missed and remove them.

This won't be fun, but *everyone* in the household must be treated when one member of the family has head lice. Lice don't care which head of hair they live in, and the infestation can last longer than necessary when family members spread lice back and forth. During the elimination process, everyone must comb their dry hair daily to check for lice. Also vacuum upholstered furniture daily, change pillowcases and bedsheets, and launder clothes. As an extra precaution, you might want to run pillows and blankets through your clothes drier to kill lice.

Check after 7 to 10 days to see if any missed lice nits have hatched; if so, you have to go through the whole process again.

AS A LAST RESORT . . .

If you absolutely *must* use chemical pesticides and you feel there is no other alternative, follow these "Pesticide Safety Tips" recommended by the Environmental Protection Agency:

- Always read the label before buying or using pesticides. Use pesticides only for the purpose(s) listed and in the manner directed.
- Do not apply more than the specified amount of the pesticide. Overdoses can harm you and the environment.
- Keep pesticides away from food and dishes.
- Keep children and pets away from pesticides and sprayed areas.
- Do not smoke while spraying.
- Avoid inhalation of pesticides.

- If you mix pesticides, do it carefully to avoid splashing. [I wouldn't mix pesticides at all! Unless you are a chemist, how could you know what toxic horror you might be creating?]
- Avoid breaks or spills of pesticide containers.
- If you spill a pesticide on your skin or on your clothing, wash with soap and water and change your clothing immediately.
- Store pesticides under lock in the original containers with proper labels. Never transfer a pesticide to a container, such as a soft drink bottle, that would attract children.
- Dispose of empty containers safely. Wrap single containers of home-use products in several layers of newspaper, tie securely, and place in a covered trash can. Never burn boxes or sacks. Dispose of large quantities in special incinerators or special landfills. [Call your local EPA office to find out where.]
- Wash with soap and water after using pesticides, and launder your clothes before wearing them again.
- If someone swallows a pesticide, check the label for first aid treatment. Call or go to the doctor or the hospital immediately and keep the pesticide label with you. [If my child swallowed a pesticide, I'd probably race to the emergency room, and not call first.]

CHAPTER 4

Tap Water ᗡ 👁 ✍ 1

It used to be that we could take for granted turning on our tap and having clean water pour out. Not anymore. Today, the responsibility for clean water lies with us. Buying the proper equipment will be a major investment, but not any more than buying any other major appliance, and it is well worth every penny for its contribution to your family's health.

We are exposed to more water pollutants than just those we drink. According to a study done by the Massachusetts Department of Environmental Quality Engineering, published in the *American Journal of Public Health,* 29 to 46 percent of water pollutant exposure (depending on the chemical and the concentration) occurs *through the skin* in children, and 50 to 70 percent occurs through the skin in adults! In addition, the EPA has now identified chloroform released from hot, running shower water as an *air pollutant*—so there's a hazard in breathing water, too.

The EPA has identified more than 700 pollutants that occur regularly in drinking water, both from municipal sources, and from water taken directly from the earth through wells or springs. At least 22 are known to cause cancer; it is not known exactly how many pollutants are carcinogenic, since not all have been tested.

It has been estimated that the 700 known pollutants may represent as little as 10 percent of the actual number of contaminants that may be present in municipal drinking waters; testing procedures have not yet been invented to detect all the dangerous substances that could be present.

The most common water pollutants are the **trihalomethanes,** or THMs, the most common of which is **chloroform.** THMs are formed when **chlorine** used as a disinfectant combines with the natural organic matter in the water (dead leaves and humus in soil, silt, and mud). According to the EPA, THMs are present in virtually all chlorinated water supplies in the United States. Chloroform can cause liver and kidney damage and central nervous system depression, and is a suspected human carcinogen.

The second most common water pollutant is **fluoride.** It is added to many water supplies as a public service, to reduce dental cavities in children. Much controversy has surrounded the use of fluoride, because in large amounts it can weaken the immune system and cause heart disease, genetic damage, birth defects, and cancer. The U.S. Public Health Service recommends an "optimum" intake of 1 mg of fluoride per day, but that level is frequently exceeded when you combine the fluoride found in tap water with the fluoride in toothpaste and mouthwashes, and the fluoridated water used in reconstituted fruit juices and other processed foods and beverages.

Pipes used to transport water throughout the system and within your home can also contribute pollutants. **Cadmium, copper, iron, lead,** and **zinc** can leach into the water from metal pipes. Cadmium can cause kidney damage, anemia, heart problems, high blood pressure, birth defects, and cancer; lead can be responsible for headaches, nerve problems, mental retardation and learning disabilities in children, birth defects, and, possibly, cancer. Asbestos cement pipes can release carcinogenic **asbestos.** The plastics industry insists that polyvinyl chloride (PVC) pipes are safe, yet studies show that a variety of toxic and carcinogenic substances leach from these pipes into the water, among them **methyl ethyl ketone (MEK), dimethylformamide (DMF), cyclohexanone (CH), tetrahydrofuran (THF),** carbon tetrachloride, tetrachloroethene, trichloroethane, **di-(2-ethylhexyl) phthalate (DEHP),** and **dibutyl phthalate.** Water standing in PVC pipes for any length of time will also become contaminated with carcinogenic **vinyl chloride.**

This list of pollutants does not even begin to cover what else might be in your particular water supply. Water is sometimes referred to as "universal solvent," because it picks up a bit of everything that it passes by. Both our surface waters and groundwaters are now contaminated from years of industrial dumping, and your water supply could contain anything from **nitrates** to **pesticides** to **industrial solvents.**

The EPA has complete responsibility for the quality of water that comes from your tap—if you get your water from a source that services more than 15 individual year-round hookups (if your source is the well in your backyard, you're not covered). All chemical additives—whether they are added for a specific purpose, or they find their way into the water supply accidentally—fall under its jurisdiction. The quality of our public water is controlled by the National Interim Primary Drinking Water Regulations (Code of Federal Regulations Title 40, Part 141), developed by the EPA in accordance with the Safe Water Drinking Act, an amendment to the Public Health Service Act. By this law, the EPA monitors *eight* inorganic chemicals and *ten* organic chemicals, leaving an estimated *30,000* possible hazardous pollutants without regulation.

I know these statistics sound incredible. It seems logical for us to expect that the EPA could do a better job of providing clean water in our taps. But we must remember that most municipal water treatment facilities were built in the early 1900s for the purpose of disinfecting water, rather than purifying it. Our water treatment facilities were designed to kill disease-causing bacteria, and not to remove chemical contaminants. The federal government estimates that it would take billions of dollars and ten to fifteen years to upgrade our water treatment facilities. Clearly, we must take the responsibility ourselves for pure water. Right now, it's the only option that is economically feasible.

WHAT YOU CAN DO

Your two choices are (1) bottled water and (2) a water filter.

Bottled water can generally be relied upon to be bacteriologically safe and free of chlorine. Many consumers believe that bottled waters are of higher quality than tap water, although legally this need not be true. Many bottled waters are only processed tap water, and

57

their quality varies widely depending on the quality of the local water and the type of filtration used. Buying "pure" water in a plastic bottle defeats the purpose because the plastic quickly leaches into the water. Federal and state regulations for bottled waters are confusing and changeable; and some bottled waters are even exempt from regulation! I drank bottled water for several years, at great expense, until I found out that the fluoride levels in my favorite brand were five times higher than fluoride levels in tap water! This was allowed because water bottlers *assume that bottled waters are not being consumed as the primary source of drinking water,* and therefore do not consider excessive pollutant levels a hazard. When I saw how unreliable bottled water can be, even though I was loyal to one brand, I decided it was time to purify my water supply myself.

Buying a water filter requires a little homework and careful consideration. There is no one filter right for every water supply and every family.

First you will need to find out which pollutants are in your water and need to be removed. Because each water purification method removes different pollutants, you have to buy the right type of equipment for your water supply.

To find out what's in *your* water, start by calling city hall, your local water district office, or your local department of health services. Be prepared for them to give you the runaround. You might not get any information from them, or they may send you the results of tests done five years ago.

You may want to ask if your water is tested on a regular basis. The Safe Water Drinking Act requires that the water be tested periodically; the interval between tests depends on the number of customers served by your water system. A survey conducted by the Congressional General Accounting Office found, however, that more than half the water suppliers were not doing the required testing. Also, remember they are only testing for eight inorganic chemicals and ten organic chemicals—there may be many other contaminants in your water. Moreover, water quality changes constantly, and pollutants that may not have shown up on the last water analysis report may be present on the next one. Some people choose to have their water tested independently, although this can be expensive and is not always accurate.

Even without test results, you can learn a great deal about the condition of your water by asking the following questions:

1. Where does your water come from? a reservoir? groundwater?

2. What type of pipe is used to transport the water? Does it add lead, asbestos, or vinyl chloride? (You also need to find out what type of pipe has been installed in your home).

3. Is chlorine used to disinfect the water? or chloramine? or some other chemical?

4. Is your water fluoridated?

5. Is there agriculture in your area that would result in excessive pesticide or nitrogen fertilizer runoff? or are there factories nearby producing industrial waste?

6. Where is your city dump in relation to municipal water supplies? Could hazardous materials from the dump be leaching into your water?

Considering the current condition of our water supplies, it is my feeling that *every* household in America needs to be concerned about the quality of water coming from their tap, and everyone must take responsibility for its purity.

For the purpose of removal, pollutants fall into four groups:

PARTICULATES (minute bits of material that do not dissolve in water)—asbestos, arsenic, heavy metals (aluminum, cobalt, chromium, nickel, mercury, lead, cadmium, manganese, silver), rust, dirt, sediment, etc.

DISSOLVED SOLIDS (solid materials that decompose in water)—fluoride, nitrates, sulfates, salts, etc.

VOLATILE CHEMICALS (nonparticulate substances that evaporate)—chlorine, chloramine, chloroform and THMs, chlorinated hydrocarbons, pesticides (DDT, dialdrin, lindane, heptachlor), PCBs, benzene, carbon tetrachloride, trichloroethylene, xylene, toluene, etc.

MICROORGANISMS (microscopic plant and animal life)— bacteria, viruses, etc.

There are three basic methods of water purification, and water purification units use these methods either singly, or in combination. The three types are activated carbon (in granular or block form), reverse osmosis, and distillation. Even though there are different brands and different designs available, each method can only remove certain pollutants, and not others.

If you know you have asbestos in your water, buying a unit to remove asbestos will, in addition, automatically remove arsenic, lead, cadmium, dirt, and all other particulates. If you want to remove fluoride, all other dissolved solids will be removed also. A unit that removes chlorine and trihalomethanes will take care of any other volatile chemicals that may be in the water.

So here's what to buy:

To remove PARTICULATES or DISSOLVED SOLIDS—distillation or reverse osmosis
To remove VOLATILE CHEMICALS—activated carbon
To remove MICROORGANISMS—distillation

For all-around protection (better safe than sorry!) you'll want to invest in either a reverse osmosis unit or a distillation unit that also includes an activated carbon filter. There are advantages and disadvantages to each, so I'll discuss them separately to help you decide which is best for you.

Activated Carbon

The least expensive water purification method is activated carbon. Even though it is totally ineffective on particulates, dissolved solids, and microorganisms, it is the best at removing volatile chemicals. So if you can't afford anything else, it is better to get an activated carbon filter than to do nothing—*if* it's the right type.

Activated carbon works by the process of adsorption. Each little particle of carbon is like a honeycomb containing minute pores that attracts and traps pollutant molecules. As water passes through and the micropores are filled, fewer and fewer pores remain available, until finally the carbon is fully saturated and it begins to release the pollutants back into the water! The game you play with activated carbon is "change the filter in time," and it can be difficult to tell exactly when to change it.

The amount of pollutants that can be adsorbed by a filter cartridge containing activated carbon depends simply on the amount of carbon in the cartridge. The more carbon, the more micropores available. The little carbon filters that you screw onto your faucet are practically worthless. They're just as good as any other carbon filter for about the first glass of water, but they become saturated very quickly and then cause more harm than good.

Another common problem with carbon filters is bacteria growth. Bacteria tend to grow inside carbon cartridges and can multiply to potentially hazardous levels. While there is no evidence that these bacteria are a common cause of disease in users of water with carbon filters, there is no practical way to monitor the amount or type of bacteria that may be present in home filters. Some granular carbon units offer a "backwashing" function to control bacteria growth, which involves switching a lever or attaching a separate device to flush hot water through the carbon in the reverse direction. Although backwashing will remove some bacteria, it is not as effective or reliable as changing the carbon completely. "Bacteriostatic" filters have been embedded with silver to control bacteria growth. EPA studies, however, have found that, in addition to being potentially harmful to health, the silver does not actually reduce the bacteria count.

Still, activated carbon filters can be effective when used properly. Carbon block units are generally preferred—the carbon lasts longer because there is more of it, and the compressed design inhibits bacteria growth. The least expensive brands come with a hard plastic housing for the cartridge. Though this does not seem to affect the water quality if the water comes in contact only briefly with the plastic while passing through, water sitting in the filter over a long period of time may absorb polymers from the plastic. Plastic units are fine for most people, but if you are extremely sensitive to plastic, you might want to get an all-stainless-steel unit.

Carbon blocks come in two basic styles: over-the-counter units that sit on top of the counter with diverter tubes to and from the tap, and undersink models mounted to the pipes that dispense filtered water through the main tap, or through an additional tap especially mounted for that purpose.

If you keep ahead of the carbon changes, you should have no problem with your filter. How often you change it really depends on the relationship between the amount of carbon in the filter and the amount of water flowing through. Try to figure out how much water you use on a daily basis and compare that to the estimated life of the carbon cartridge. Mark the date on your calendar and change the carbon when that time has passed, even if you think it is unnecessary. It doesn't hurt to change the carbon too often. At the very least, change it once a year.

There are many carbon block filters on the market. Compare them according to the price of the original unit and the price of the cartridges *and their estimated life.* I did a cost comparison once and there was a difference of *more than $200* between the least and most expensive units to filter the same amount of water.

There are two special types of granulated carbon filters you might want to consider—one is for the shower head and the other filters water for the whole house. Because skin absorption of water pollutants and inhalation of volatile chemical fumes during showering has now been proved to be a problem, I strongly suggest that you do something to protect yourself from these other modes of exposure. A whole-house filter is attached to the incoming cold water line of your home, so all water is filtered, even in the toilet. You have clean water for bathing, teeth-brushing, and dishwashing; all taps dispense clean water. A shower-head filter screws right into your shower head and filters only the water at that point. I know many people who have these simple shower-head filters and I have heard some amazing stories. One friend used to have a horrible rash all over her body that required cortisone treatments. Soon after she began filtering her shower water, the rash went away completely and has not returned. If you have a skin problem, you really may want to try this. I have had one on my shower head for years and I love it.

Reverse Osmosis

Reverse osmosis purifies water by forcing it through a membrane that allows water molecules through, but not pollutants. Plants purify water by a similar method—as the water passes through the natural cellulose, nutrients are removed and waste fluids disposed. Reverse osmosis is a very good method of removing particulates and dissolved solids, but will not remove microorganisms, and barely touches volatile chemicals. For this reason, most reverse osmosis membranes are used in conjunction with a carbon filter. If particulates and dissolved solids are not a problem in your water supply, the added expense of a reverse osmosis membrane is unnecessary.

A reverse osmosis unit sits under the sink and is easy to use and maintain. All you have to do, once it's installed, is flip the lever on a specially installed tap, and out comes clean water. Periodically you have to change the membrane, the carbon, and the special particulate prefilter, but otherwise there is no involvement with the process.

These units do have certain drawbacks, however. The biggest disadvantage is that the quality of water diminishes with use, and filtration many be inconsistent, depending on water pressure. At best, while it is more effective than just plain carbon, for overall pollutant removal reverse osmosis just can't match the performance of a distiller. And since they are both in the same price range, why settle for second best?

Distillers

For high quality, consistently pure water, nothing beats a distiller. Distillation comes closest to duplicating nature's own hydrologic cycle for water purification. The heat of the sun causes water to evaporate from the earth's surface, leaving impurities behind, then it gets condensed, and pure water returns to the earth as rain, hail, sleet, or snow.

Water distillers work by boiling the water to turn it into steam and then condensing it into "pure" water. Boiling the water destroys bacteria and other living materials, leaving them behind in the boiling tank, along with particulates and dissolved solids that are too heavy to rise with the water vapor.

Early distillers concentrated only on the removal of microorganisms and solid materials. Newer designs have special volatile gas vents or double-distilling processes that enhance volatile chemical removal. Most are also coupled with activated carbon postfilters to remove any chemicals that may remain after distillation.

One danger with stainless steel distillers is that they add aluminum to the water. In a laboratory test done by the Rodale Press Product Testing Department, all of the metal distillers tested produced water with at least traces of aluminum; some distillers increased the aluminum content of the water by up to 130 percent, bringing the aluminum levels to the limits generally considered acceptable. Because aluminum exposure has been linked with nervous system diseases and brain disorders, the fact that stainless steel distillers add this particular metal to the water is cause for concern. You can avoid this problem by choosing a glass distiller.

The major disadvantage to distillers is that you have to "make" the water. It's not automatic, you usually have to turn on the power supply and hook up the hoses to the water source. I can tell you from experience, this is not a big deal. I have a distiller mounted near my kitchen sink. It's as simple to turn on as turning on a light bulb, and a flick of a finger snaps the input water hose to the faucet. I can't use my kitchen sink for a few hours, but it's a small price to pay for consistently pure water, gallon after gallon, for the rest of my life.

Where to Buy Water Filters

Now that you have some idea of what's available, where are you going to buy it? Not at your local hardware store. You probably will have to order what you need by mail (see Resources), although you might ask your natural food store if they know of any local dealers.

CHAPTER 5

Drugs and Medications ☞

Most home medicine cabinets are filled with over-the-counter, nonprescription drugs—pain killers, antacids, allergy medicines, cough syrup, laxatives, sleeping pills—it's a multi*billion* dollar industry offering more than 200,000 products.

An over-the-counter drug, as opposed to a prescription drug, is a drug product you can buy without a prescription and use when *you* think it is useful or necessary, without a doctor's guidance. Over-the-counter drugs are formulated only to relieve symptoms and not to cure the underlying disease. They just make you less miserable so you can continue your daily routine while your body works on curing itself with its natural healing powers.

The biggest danger in having any kind of drug or medication in your home is the risk of accidental poisoning from an overdose.

Children are attracted to the brightly colored pills and can quickly consume too many, with serious consequences.

Many drugs and medications have dangerous side effects. I'll be discussing the specific risks associated with some of the more common nonprescription drugs later in this chapter, but if I don't cover everything in *your* medicine cabinet, you can go to your local library and look up information on over 2,500 popular pharmaceuticals in the *Physician's Desk Reference.* There you'll also find other books written for the general public about the health effects of both nonprescription and prescription drugs. Your favorite bookstore will also carry books on the subject. You should do this extra research because many times the side effects of these products are not listed on the product label.

Only active ingredients are required to be listed on the labels of drugs and medications, and they probably contain many additional ingredients you might want to watch out for: **alcohol, caffeine, artificial colors and flavors, sugar, saccharin, preservatives,** the list is endless, and there is no way of knowing if these additives are in the product.

In using over-the-counter medications, we are, in effect, playing doctor with ourselves, and, as our own self-care physicians, we want to prescribe the proper treatment. Yet, how often do we take medications, or give them to other members of our household, without stopping first to find out *all* the facts—the proper dosage, possible side effects, and other health effects, or if in fact this is the right drug for the symptoms. Since 1962 the Food and Drug Administration has been reviewing the active ingredients found in over-the-counter drugs. In many cases they have found that, while some active ingredients may be "safe," they are not effective at doing what the consumer believes the drug is supposed to do. With some products, we are taking medications we don't even need to be taking because we are misdiagnosing our symptoms!

Symptoms are signals from our body that something is wrong, and that our body is doing something about it. When we run a fever, our body is heating up to kill germs; when we cough, it is to clear our lungs; a headache may be a sign we are under too much stress and should take more leisure time for ourselves. It's not always a good idea to stifle our symptoms, and never a good idea to ignore them completely, as they are part of our natural healing process.

Over-the-counter medications are meant to provide *temporary* relief for *occasional* symptoms, and are not to be used on a regular basis. As the label says, "If symptoms persist, see your doctor." You may have a more serious illness that requires medical attention.

Because all you are doing with an over-the-counter drug is making yourself more comfortable, why not alleviate your symptoms without using drugs?

ARTHRITIS, FEVER, ACHES, AND PAINS

The most common reason people take nonprescription drugs is for pain relief. One evening spent watching television will verify this point, since your favorite shows will be interrupted every ten minutes by commercials for Bayer aspirin, Anacin-3, Extra-Strength Excedrin, Advanced Formula Dristan, Advil, Nuprin, Tylenol, and Vanquish.

Most pain relievers also lower fevers and reduce the inflammatory distress of arthritis, and these drugs are frequently advertised to relieve all these symptoms.

Painkillers, or analgesics, work by blocking the transmission of pain impulses to the brain and should not be taken for more than ten days in succession by adults, or for more than five consecutive days by children.

Most analgesics contain one of three active ingredients: aspirin, acetaminophen, or ibuprofen.

Aspirin is by far the most widely used painkiller. Over 20 billion doses are sold each year; that's about 100 aspirin tablets for every man, woman, and child in America. The FDA considers aspirin to be "safe and effective" when taken in the recommended dosage, for no longer than 10 days and if no more than 4000 mg is taken in a 24-hour period. Aspirin has been widely used for years and is considered safe for most people; however, in some cases aspirin can

■ interfere with blood-clotting, and regular use can cause iron-deficiency anemia;

- trigger or aggravate peptic ulcer and cause stomach upset, heartburn, or bleeding in the stomach;
- produce allergic reactions in an estimated 2 out of 1000 people and in some asthma sufferers. Reactions can range from rashes, hives, and swelling, to life-threatening asthma attacks;
- stimulate the brain and spinal cord, followed by central-nervous system depression, which appears as respiratory failure, circulatory collapse, coma, and death;
- in high doses, interfere with liver function.

Early signs of aspirin toxicity include ringing in the ears, hearing loss, *headache,* dizziness, vomiting, rapid breathing, extreme irritability, and bizarre behavior. Pregnant women should check with their physicians before taking aspirin. Children under age 18 with chicken pox or flu should not be given aspirin, as they may develop Reye's Syndrome, an often fatal liver disorder.

The FDA also considers acetaminophen to be "safe and effective" when taken in the recommended dosage, for no longer than 10 days and if no more than 4000 mg is taken in a 24-hour period. Milligram for milligram, acetaminophen is as effective as aspirin for pain relief and fever reduction, but is not recommended to reduce the inflammatory effects of arthritis.

Acetaminophen is used by many people who are allergic to or who get upset stomachs from taking aspirin. For this reason, it appears to be safer than aspirin, but in some ways it's not. You can be allergic to acetaminophen too, and overdoses can cause severe liver damage, even death.

Ibuprofen is a relative newcomer to the painkiller market, having been available for many years only as a prescription drug. Possible side effects include nausea, vomiting, stomach cramps or pain, constipation, diarrhea, heartburn, stiff neck, headache, fever, dizziness, depression, insomnia, blurred vision, or swelling of hands and legs. You should not take ibuprofen if you are allergic to aspirin or have a history of dizziness, bronchospasms, liver disease, hypertension or heart disease, nasal polyps, stomach ulcer, or intestinal bleeding. According to *Physician's Desk Reference,* "It is especially important not to use ibuprofen during the last three months of pregnancy unless specifically directed to do so by a doctor because it may cause problems in the unborn child or complications during delivery."

WHAT YOU CAN DO

Let's start with headaches, and some amazing statistics. Approximately 9 out of 10 headaches are the result of apprehension, anxiety, depression, worry, and other emotional states. FDA investigators found that over-the-counter pain relievers provide *little if any relief for this type of headache.* Other common types of headaches are migraine headaches and hypertensive headaches, caused by a sudden rise in blood pressure. Authorities say that *these types of headaches should not be treated with over-the-counter analgesics either.* The other types of headaches are caused by inflammation—of the sinuses, of membranes surrounding the brain, or even from developing brain tumors. For these conditions, *medical care is recommended, not over-the-counter drugs.* So when is it appropriate to take an analgesic for a headache? If it's caused by a fever or a hangover; otherwise the FDA says no.

Headaches can be caused by a great variety of food allergies or by many substances you might have around the home. If headaches are a recurring problem for you, it might benefit you greatly to play detective and keep a diary of everything you eat, every exposure to a chemical, every activity, and every time you get a headache. This seems like a lot of trouble, I know, but it's very rewarding when you no longer have headaches because you've found the cause, instead of just covering up the effects.

For temporary relief from occasional headaches, regardless of the source, drink a cup of strong peppermint tea and take a short nap. Or try teas made from rosemary, catnip, or sage.

Massage can also help a great deal. Gently rub the area where the pain is, or relax your neck by letting your head droop forward as far as it will go, then turning your head slowly in the widest possible circle. Lying in a warm bath with a cold washcloth on your head will lessen pain by drawing blood away from your head. And, finally, I have a friend who swears that the best cure for his rare headaches is a steaming bowl of Chinese hot and sour soup, without MSG. It works every time.

A fever is generally considered to be any body temperature over the average 98.6° F. Any chronic or extreme symptoms and temperature over 102° F warrants immediate medical attention. But for those in-between fevers, there are a few things you can do.

First, decide if you really want to lower the fever. In many cases, fever can be beneficial, as it seems to increase the mobility of your white blood cells and enhances their ability to kill germs. One problem with taking aspirin to reduce fever is that it interferes with the body's natural defense cycle, so while you may reduce your fever, it might take longer for you to get well.

The simplest way to reduce fever is by placing a cold washcloth on the forehead. It has to be changed and rinsed regularly, as it quickly absorbs heat from the skin. For a serious fever, overall sponging of the body, taking a cool bath or shower, or wrapping the patient in a wet sheet for short periods of time will help.

Various herbal teas have been used to treat fever, including boneset, yarrow, vervain, and barberry berries. Your natural food store may carry these herbs, or see if you can find an herb shop in your community. If it's too much trouble to go hunting for these herbs, open your spice cabinet and take out your cayenne pepper—yes, *hot* red pepper. Add small amounts to warm water, milk, or tea, or fill empty gelatin capsules and take a couple with a glass of water.

You can get dehydrated very quickly with a fever, so keep drinking liquids. Add as much lemon juice as you can to everything and take some extra vitamin C.

HEARTBURN, CONSTIPATION, DIARRHEA, AND NAUSEA

Between the television commercials for painkillers are the commercials for antacids—liquids, gels, tablets, capsules, powders—to relieve the "heartburn," "sour stomach," and "acid indigestion" caused by "excess stomach acid." It seems there's something wrong with everybody's stomachs—if it's not heartburn, it's "irregularity." Diarrhea isn't quite so popular in the TV commercials, but it's all part of the same gastrointestinal problem. Nausea is sometimes related to gastrointestinal function, but can also be a sign of food poisoning, hormonal changes (during pregnancy or menstrual cycle), a food allergy, or motion sickness.

Antacids work by chemically neutralizing excess hydrochloric acid in the stomach. They are considered to be relatively safe, in comparison with other groups of drugs; however, they do contain ingredients that can be dangerous to some people, particularly if they are taken in large amounts over a long period of time.

Some antacids contain aluminum compounds, which can be risky for people with kidney problems or who are on kidney dialysis. The level of aluminum in your blood doubles when you take an aluminum-containing antacid. Heavy use may affect your body metabolism in its ability to process certain essential minerals, which may lead to bone abnormalities. Aluminum can also contribute to the development of Alzheimer's disease, a form of early senility.

Bicarbonate compounds are also widely used as antacids, including sodium bicarbonate (baking soda) and potassium bicarbonate. Because they are readily absorbed into the body, they can measurably increase the alkalinity of blood plasma and other body tissues. Sodium bicarbonate also increases the sodium level in the body, which may be risky for those with high blood pressure.

Most laxatives are not especially harmful, but the FDA believes that there is a widespread overuse of laxatives, due largely to misleading advertising about "irregularity." Laxatives should be used *only* in cases of infrequent, difficult, or uncomfortable bowel movements, as an occasional temporary measure, and not on a regular basis, or for an extended period of time. Prolonged use of laxatives can seriously impair normal bowel function, and people can become dependent on laxatives. It is normal for bowel movements to occur in the range of 3 movements per day to 3 movements per week, and any sudden change in your normal pattern that lasts for more than a couple of weeks is cause for you to seek professional help, not treat yourself with a laxative.

The FDA also frowns on diarrhea remedies. Over-the-counter products are effective for only the mildest forms of diarrhea, which usually go away by themselves in a day or two. One of the most popular antidiarrheal drugs, **diphenoxylate hydrochloride**, may cause more harm than it does good. It is a narcotic substance that can cause addictive symptoms, and an overdose could be fatal. To prevent overdose, the drug is formulated with other substances that can cause dry skin, flushed face, rapid heartbeat, and other unpleasant side effects.

WHAT YOU CAN DO

All gastrointestinal difficulties can be related to the food you eat, so if your stomach and intestines are giving you trouble, they are probably just unhappy with what you're putting in them.

The first thing to look at are the foods you are eating. Do you eat a lot of processed foods? If so, you probably have a diet high in sugar and fat, and low in dietary fiber, which helps the gastrointestinal tract work smoothly. Are you drinking enough water and other beverages? A diet that contains lots of fresh fruits, vegetables, and whole grains provides adequate amounts of fiber and fluids to keep your gastrointestinal tract well-regulated.

If you occasionally have heartburn and need an antacid, don't reach for an over-the-counter drug. An interesting study in Great Britain showed that antacid tablets containing the usual aluminum hydroxide, magnesium, and sodium bicarbonate didn't work at all to reduce heartburn. What did work for about half the patients was a formula that contained alginic acid, a natural substance derived from seaweed. If you'd like to try this, get some sodium alginate tablets from your natural food store, and chew one or two tablets thoroughly (do not swallow them whole), and then drink a glass of milk.

Losing weight can also help heartburn, as can eating smaller meals, especially in the evening, and avoiding lying down for several hours after the meal. Alcoholic beverages, chocolate, coffee, tomatoes and tomato products, citrus fruits, and fried, fatty, or spicy foods all stimulate your stomach to produce excess acid. Cigarette smoking also greatly contributes to heartburn—if you smoke, there's a good chance that by giving up your cigarettes you'll also be giving up your heartburn.

For occasional constipation, try prunes or prune juice or large doses of vitamin C. Your natural food store probably carries some herbal laxatives that are very effective, in addition to psyllium seed, another natural substance, FDA-approved as a safe and effective laxative.

Diarrhea, like fever, is one of the body's healing mechanisms —if you get it, it probably means that your body wants to expel whatever is in your gut. It may want to get rid of toxic virus or bacteria or cleanse your system of a mild case of food poisoning. Or, you might have an allergy to the food the body wants to eliminate. So it might not be in your best interest to inhibit diarrhea, but again, don't ignore chronic or severe symptoms.

To relieve nausea, try a cup of raspberry leaf or basil tea. During pregnancy, 10 to 20 mg per day of vitamin B6 has been known to help.

My favorite remedy for nausea is ginger ale. I had a medical doctor once who was very old and knew a lot of folk remedies. He told me to drink ginger ale—ginger ale specifically, and not other carbonated beverages—because the ginger helped the nausea. Instead of drinking commercial ginger ale, I make my own with fresh ginger and honey. Here's the recipe (I know it's a little complex and not what you want to do when you're sick, but you can make the syrup and keep it indefinitely in the refrigerator, if you can resist drinking it in the meantime!):

NAUSEA RELIEF GINGER ALE

6   ounces fresh gingerroot (available at most supermarkets)
2   cups water
1¼ cups honey

1. *Peel and finely chop the ginger (you should have about one cup). If you have a food processor, it's very easy.*

2. *In an enamel or stainless steel saucepan, bring the ginger and water to a boil, then simmer for 5 minutes. Remove from heat and let stand for 24 hours, covered with a cloth (a kitchen towel works fine).*

3. *Strain through two layers of cheesecloth (buy at hardware store), and squeeze the pulp in the cloth to extract all possible juice.*

73

4. *Return juice to saucepan, add honey, and bring to a boil over moderate heat, stirring to dissolve the honey. Simmer for 5 minutes.*

5. *Cool, pour into a bottle and refrigerate.*

TO USE: Mix a soupspoon of syrup in a glass of carbonated mineral water or club soda, more or less to taste.

ALLERGIC HAY FEVER

Each year millions of allergy sufferers take antihistamines to relieve their symptoms. During an allergic response, the body produces an irritating substance called histamine, which attaches itself to the secretory cells of the nose, eyes, and lungs. Antihistamines block the histamine receptor sites on these cells, preventing the condition that causes the symptoms.

In general, antihistamines are considered to be safe and effective. There are only two notable side effects: dry mouth and drowsiness. If you'd like to stay alert and be symptom-free, try a natural remedy.

WHAT YOU CAN DO

A basic, all-purpose hay-fever remedy is to take 500 mg of vitamin C three times a day, and twice a day take 50 mg of pantothenic acid and a teaspoon of grated orange or lemon rind sweetened with some honey. You might not get immediate relief, but chances are you should notice a marked improvement within three days.

Why does this work? Vitamin C is a natural antihistamine and the bioflavinoids in the citrus peel enhance its bioavailability. Pantothenic acid alone has been known to combat allergy symptoms, and up to 500 mg per day can be taken for this purpose. Experiment for yourself with these substances and you will find just the right combination to keep you from reaching for the antihistamine bottle.

Another solution for hay fever is to figure out what is causing it, and remove the item, or items from your home. This will work if you are allergic to your cat, or the feathers in your pillow, but it might be more difficult to control if you are allergic to the acacia outside your bedroom window.

COLDS AND COUGHS

Cold and cough medications contain a complex variety of active ingredients required to relieve the aches, sore throat, coughs, fever, nasal and sinus congestion, runny nose, and sneezing typical of the common cold. The FDA emphasizes that, while these products may relieve symptoms, they *cannot "cure" colds.*

Most general-purpose cold medications include cough suppressors and expectorants, nasal decongestants, sore throat medications, and pain killers/fever reducers. Each of these drugs can also be purchased as a separate product to be taken for specific symptoms.

Cough suppressors temporarily inhibit the impulse to cough. Codeine and dextromethorphan act in the brain to depress the activity of the cough center; other drugs alleviate pain and irritation in the throat and bronchial passages to lessen the need to cough. Over-the-counter cough suppressors are intended to be used for periods of less than one week to diminish coughs due to bronchial irritation. People who have asthma, emphysema, and other lung conditions should not take over-the-counter cough suppressors for relief; neither should those who smoke. The primary danger of cough supressors that contain codeine is the potential for addiction, and in many states they cannot be purchased without a prescription. Dextromethorphan can cause stomach upsets, drowsiness, and unconsciousness. Some cough syrups contain very high levels of alcohol; a few preparations contain even more alcohol than some alcoholic beverages. Expectorants are also found in cough preparations, but the FDA has not approved any as being safe and effective.

Nasal decongestants constrict swollen blood vessels in the lining of the nose and sinuses, which are the primary cause of stuffy noses. Unfortunately, when the drug wears off, the blood vessels may swell even more than before, making your stuffy nose worse! Decongestants should be used no more frequently than 3 days in a row, and not without doctor's orders if you have high blood pressure, heart disease, diabetes, or thyroid disease.

Sore throat medications are designed to provide only temporary relief of minor symptoms and should not be used for more than 2 consecutive days. Most sore throats are a symptom of an underlying infection or illness that should be treated, not hidden under a cough drop.

WHAT YOU CAN DO

Of course, the first cold remedy that comes to mind is vitamin C, right? Although there is little scientific evidence to show that vitamin C *prevents* colds, it can significantly shorten the cycle of the cold virus, even if you wait until the first sign of a cold to start taking the vitamin. Start by taking several 500-mg doses throughout the day *as soon as you notice the symptoms.* You can take up to 4,000 or 5,000 mg each day for the duration of your cold, if you need to.

Also, take lots of hot fluids, chicken soup, and anything spicy. Cayenne pepper and garlic are both germ killers. Among the herbal teas, try chamomile, lemon balm, boneset, elder, or vervain.

To soothe the tickle in your throat, almost any hard candy will work as well as a cough drop. The idea is just to keep you salivating. Your natural food store might have some herbal drops or honey-sweetened hard candies.

When I was a child and got really bad sore throats, my parents would make a horrible-tasting concoction of honey and onions. They would slice a whole white onion in a bowl, cover it with lots of honey, then cover the bowl and let it sit. After a few hours, the onion would "sweat" and release its juice. I was given a couple tablespoons of this honey/onion juice and by the next day I was magically cured.

You could also try hot ginger milk. Heat milk and add two or three slices fresh ginger (or ½ to ¾ teaspoon ground ginger) and honey to taste. Serve hot.

INSOMNIA

Because many people have difficulty falling asleep or staying asleep, the FDA sees a valid need for safe and effective nonprescription sleep aids. However useful and convenient such a product might be, the FDA has found that most active ingredients used in sleeping pills are unsafe, ineffective, or both.

For many years the most popular active ingredient in sleep aids was the antihistamine methapyrilene. In 1979, products containing methapyrilene were recalled when the National Cancer Institute found that methapyrilene caused cancer in laboratory animals and posed a potential human hazard.

Over-the-counter sleeping pills are intended to be used no longer than 2 weeks. Chronic insomnia warrants professional attention.

WHAT YOU CAN DO

First of all, you may not even have insomnia. A good number of "insomniacs" are only insomniacs because they think they need eight hours of sleep every night. Our bodies vary from one individual to another, and we each require different amounts of sleep. We need only enough sleep to feel rested, and that amount may change at different times in our lives.

The best "medicine" for a good night's sleep is a mug of warm milk. Milk contains the amino-acid tryptophan, which aids your body in the production of sleep-inducing serotonin. Also, a high-carbohydrate evening meal will help you easily drift off to sleep.

Chamomile tea has been shown in scientific studies to have a sleep-inducing effect on 10 out of 12 subjects. Other herbs known to have a similar effect are hops, passionflower, catnip, basil, and lemon verbena.

As a former insomniac myself, I can tell you a couple of things that worked for me. First, I changed the sheets on my bed from polyester/cotton to 100 percent cotton. This worked like magic—the very first night I went right to sleep. I later found out that polyester/cotton sheets are heavily treated with formaldehyde, which is known to cause insomnia! After I got that out of the way, I found that what was really keeping me awake was that although my body was tired, my mind wouldn't relax. It was racing along like an endless tape recorder on fast forward. So I began learning to quiet my mind by repeating the word "sleep" over and over. At first I couldn't do it because my mind would wander too much, but I just kept at it, night after night, until now I just lie down in bed, close my eyes, think "sleep," and I'm out.

IF YOU MUST TAKE MEDICATIONS . . .

By the time you have read this far, I hope you are convinced that over-the-counter drugs are not what you want to take for your symptoms, if for no other reason than that they can't cure any of your ills. But, if you still want to take them (I'm not encouraging you to) please use them with good judgment. Here are a few tips for responsible use of medications:

- Prevent accidental poisonings by keeping medications out of children's reach.
- Always check with your doctor to make sure the over-the-counter drug you have chosen is appropriate for your symptoms, *especially if you are pregnant.* Always tell your doctor *all* the medications you are taking so he or she can assess the possible dangers of mixing medications.
- Take the medication only as directed on the package.
- Do not take any more medications than are absolutely necessary, as the greater the number of drugs taken simultaneously, the greater the likelihood of adverse effects.
- Read package labels and any other information you can find on the drug you have chosen.
- Be aware of how you feel while under the influence of the drug, and watch out for side effects.

CHAPTER 6

Personal Care Products ☞ ✍ ✋

Beauty and hygiene items are applied to some of the most sensitive parts of our bodies. We use soap, shampoo, toothpaste, contraceptives, colognes, makeup, toilet paper, and other personal products day in and day out. All are applied to our skin, some in very delicate areas. You would think these products are tested for safety according to regulations as strict as those for the food we eat. Unfortunately, they are not.

Most hygiene products are regulated by the FDA as "cosmetics," a category that includes anything that can be "rubbed, poured, sprinkled or sprayed on, introduced into, or otherwise applied to the human body . . . for cleansing, beautifying, promoting attractiveness, or altering the appearance without affecting the body's structure or functions." The law does not require that cosmetics be tested for safety before they are allowed to be sold. The FDA can take action on a case in which harm is done only after a product is on the market, and only after it has received enough consumer complaints and enough evidence is collected to prove in court that the product is hazardous. *Then* the FDA can halt its production and sale.

Thousands of different ingredients derived from petrochemicals or natural animal, vegetable, or mineral sources are used in cosmetics. The FDA requires a complete listing of ingredients on all domestic cosmetics, itemized in decreasing order and expressed in standardized language. Some commonly used items that we think of as cosmetics are exempt from this labeling requirement. Deodorant soaps, fluoridated toothpastes, antiperspirants, sunscreens, and anti-dandruff shampoos all claim to affect the body's structure or function and are regulated as over-the-counter drugs.

Not surprisingly, the most common complaint associated with personal care products is skin rash, which can range in intensity from moderately irritating to painful and disfiguring. But there are also other hazards. Did you know your lipstick may be carcinogenic? Or that an accidental swallow of perfume could kill your child?

There are a number of safe beauty and hygiene products on the market, and you can make many yourself at home. It's mostly a matter of getting back to basics and using natural products that are in harmony with your body. I'm not suggesting to you women that you will have to change the way you look, though you may choose to do that when you see the gorgeous colors available in natural cosmetics, or the subtle highlights in your hair when you rinse it with camomile tea, and how softly your hair stays in place with lemon juice hair spray. A warm, fragrant herbal bath is much more luxurious than any bathtub full of bubbles. Take care of yourself with products that not only make you more attractive, but are good for you too. All the top fashion magazines and the most beautiful women in the world agree that you are the most beautiful you can be when you are healthy and fit and use beauty products to subtly enhance your own beauty, rather than drastically changing the way you look, which hides the real you.

Before we get into specific natural products, you need to know that there is no legal definition of the term *natural* when used on cosmetic items. Because natural beauty and hygiene items are becoming big business, some not-so-natural manufacturers are trying to capitalize on this market and putting the word *natural* on their labels. Their products may sound natural, with their jojoba oil, honey, herbs, and wheat-germ oil, but they may also contain such unnatural ingredients as artificial colors, fragrances, and preservatives. So you really need to read labels and find out as much as you can about the ingredients in a product before you buy it.

If you are allergic, you need to also beware of products labeled *hypoallergenic.* Because allergies are so individual, no one product can be truly hypoallergenic for everyone. *Hypoallergenic* simply means that the most *common* allergens have been removed— fragrance, lanolin, cocoa butter, cornstarch, cottonseed oils—but these products still may contain ingredients to which you are sensitive.

I was surprised to find that some hygiene products, which are formulated to use on or around the body, even in imtimate areas, actually have warning labels. We'll start with those.

DENTURE CLEANERS ☙ 👁 ✍

From the Manufacturers' Warning Labels:

> **DANGER:** Injurious to eyes. Harmful if swallowed. Keep out of reach of children.

Despite its ominous warning, denture cleaners are actually relatively safe to use. Most are made up of a combination of salts— sodium perborate, sodium chloride (table salt), magnesium sulfate (Epsom salt), calcium chloride, sodium carbonate, and trisodium phosphate—which, while not intended to be eaten, will cause no more than an upset stomach if accidentally ingested by a child.

These salts are very irritating to eyes, however, so if you leave your teeth sitting around in a glass and have little ones in the house, make sure to keep the glass out of their reach. Children are inquisitive, and they could reach for the glass and spill it into their eyes.

Denture cleaners can also irritate the skin, but again the hazard here is minor.

The only other ingredients found in denture cleaners are **artificial colors and flavors** and various **preservatives,** neither of which pose a real hazard by this method of exposure.

WHAT YOU CAN DO

Now that you know there's very little danger in using denture cleaners, you might want to just go ahead and use them at home. But you can save some money by making them at home. These recipes use exactly the same active ingredients as the commercial denture cleaners do and so will clean them just as well. But remember to take the same precautions around children because these homemade potions can also irritate their eyes.

- The simple method: dissolve ¼ teaspoon trisodium phosphate (TSP) into ½ glass of water and soak dentures overnight. This is the same TSP you may have already purchased at your hardware store to use for nontoxic cleaning. If you want something fancier, shake ⅝ cup of TSP together in a bottle with 7 drops of essential oil of cinnamon, or peppermint, or whatever flavor you prefer (available at your natural food store), then proceed as above, using ¼ teaspoon of the flavored TSP to ½ glass of water.

- The complex method: mix together 1¼ cups TSP (plain or flavored), ⅝ cup sodium perborate (a bleaching agent, available at drugstores), and ⅝ cup salt. Dissolve ¼ teaspoon in ½ glass water and soak overnight. This combination of salts is very similar to the commercial formulas and has a bleaching action that plain TSP doesn't have.

Of course, you should rinse your dentures well in the morning before wearing them.

FEMININE DEODORANT SPRAY AND DOUCHES

From the Manufacturers' Warning Labels:

> **WARNING: Avoid spraying in eyes. Contents under pressure. Flammable. Do not puncture or incinerate. Do not store at temperatures above 120° F or use near fire or flame. Keep out of reach of children. Use only as directed. Intentional misuse by deliberately concentrating or inhaling contents can be harmful or fatal.**
> **CAUTION: For external use only. Spray at least 8″ from skin. Do not apply to broken, irritated, or**

itching skin. Persistent, unusual odor or discharge may indicate conditions for which physician should be consulted. Discontinue use immediately if rash, irritation, or discomfort develops.

Feminine hygiene sprays generally contain only a pleasant fragrance; they do not contain antibacterial agents to stop odor. Irritation is common and can range from rashes and discomfort to infection and open sores. Men can also suffer from skin problems if their partner uses a deodorant spray prior to intercourse.

A long-term danger to using feminine hygiene sprays on a regular basis is that most feminine deodorants contain **talc**, which can be contaminated with carcinogenic **asbestos**. Studies have shown that there is no safe level for asbestos exposure. Asbestos causes its greatest harm when inhaled into the lungs, and when it is dispensed with an **aerosol spray**, as these products are, it is very likely to end up just where you don't want it.

Similarly, feminine douches often do more harm than good. They generally contain **ammonia, detergents, artificial fragrance, and phenol.** Phenol is a very toxic chemical that is easily absorbed by the skin. Douches can cause irritation, allergic skin reactions and "chemical vaginitis." In addition, toxic chemicals are easily absorbed by the delicate skin inside the vagina, compounding their danger. It sounds to me like a big risk to take for a product most physicians think is totally unnecessary for healthy women.

WHAT YOU CAN DO

A daily rinsing with plain water while bathing should keep the vaginal area odor-free. Do not use soap inside, as it may cause irritation for you or your sexual partner.

If you regularly have a disagreeable odor, consult your physician and check for infection, rather than trying to cover up the smell.

AEROSOL HAIR SPRAY AND STYLING MOUSSE ☞ ☜ ✑ 👁

From the Manufacturers' Warning Labels:

> WARNING: Flammable. Avoid fire, flame, or smoking during application and until hair is fully dry. Avoid spraying near eyes. Contents under pressure. Do not puncture or incinerate. Do not store at temperatures above 120° F. Keep out of reach of children. Use only as directed. Intentional misuse by deliberately concentrating and inhaling contents can be harmful or fatal.

Common ingredients in hair spray include **aerosol propellants, alcohol,** carcinogenic **polyvinylpyrrolidone plastic (PVP), formaldehyde,** and **artificial fragrance.**

Regular users of hair spray run the risk of developing a lung disease called thesaurosis, which causes enlarged lymph nodes, lung masses, and changes in blood cells. Fortunately, the disease is reversible. An FDA report noted that in one study more than half of the women afflicted with this disease recovered within six months after discontinuing hair spray use.

Also, many people have allergic skin reactions to hair spray that ends up on delicate facial skin instead of on the hair. Eye and nasal irritations are not uncommon side effects.

Styling mousse contains almost exactly the same ingredients as hair spray. The only real difference is that it comes in a foam instead of a spray.

WHAT YOU CAN DO

Check your local natural food store for an unscented natural hair spray in a nonaerosol pump spray bottle. These are much safer than the standard brands in the aerosol can, but still may contain alcohol, perfume, and other ingredients that can cause allergic reactions in some people.

You can make your own hair spray from lemons! I don't use hair spray at all, but I have many friends who use this one, and they tell me it keeps their hair in place and still leaves it feeling soft. No, it doesn't get sticky if the proportions are right, and it smells less than an aerosol hair spray. Chop 1 lemon (or one orange for dry hair). Place in a pot, cover with 2 cups pure, hot water, and boil until only half remains. Cool and strain. Place in a fine spray bottle and test on hair. If it's too sticky, add more water. Store in the refrigerator, or add 1 ounce vodka per cup of hair spray as a preservative (with the vodka you can keep this hair spray unrefrigerated for up to two weeks).

If that seems like too much trouble, you can also make hair spray with honey! Put 2 to 5 teaspoons of honey in a spray mist pump dispenser with about a cup of warm water and shake well. You'll have to experiment a bit with the proportions. The more honey it contains, the greater the holding power, but too much honey will make your hair sticky. Store the mixture in the refrigerator.

Now, if you prefer to use mousse, try gelatin instead. Dissolve ¼ teaspoon of plain, unflavored gelatin in 1 cup of boiling water, and let it sit at room temperature (do not refrigerate) until slightly set. You can make this in advance and keep it in a jar in the bathroom. To use, just rub into wet or dry hair with your fingers and blow dry. It leaves no residue but gives lots of body!

DANDRUFF SHAMPOO

From the Manufacturers' Warning Labels:

> **CAUTION: Not to be taken internally. Keep out of reach of children. Avoid getting shampoo in eyes—if this happens, rinse eyes with water. If irritation occurs, discontinue use.**

Dandruff shampoos are the most dangerous of all hair care products because they contain highly toxic medications to prevent the scalp from peeling. One popular antidandruff agent is **selenium sulfide** which, if swallowed, can cause the liver, kidneys, stomach, heart, and other organs to degenerate. Another toxic ingredient, **recorcinol,** is very easily absorbed through the skin and can lead to inflammation of the inner eyelids, skin irritation, dizziness, restlessness, rapid heartbeat, breathing difficulties, drowsiness, sweating, unconsciousness, and convulsions. In addition, dandruff shampoos may contain toxic **cresol,** carcinogenic **polyvinylpyrrolidone plastic (PVP), formaldehyde, detergents, artificial colors,** and **fragrance.**

Regular shampoos also have their risks. **Formaldehyde** is commonly used as a preservative, and may be present in shampoos hidden under the name **"quaternium-15."** In addition to being a potential human carcinogen, formaldehyde can be an irritant to skin, eyes, and respiratory passages, even at very low concentrations. Because even small, indirect exposure to formaldehyde can cause adverse health effects, government regulatory agencies require other types of products containing this chemical to carry warning labels, but because of inconsistencies in labeling laws, shampoo does not.

Another hidden and unpredictable hazard is sometimes found in shampoos—this is the intimidating sounding **2-bromo-2-nitroprone-1, 3-diol (BNPD).** This chemical can create carcinogenic nitrosamines when combined in a bottle with triethanolamine (TEA) or diethanolamine (DEA), two rather harmless ingredients found in most shampoos, or with amines on your skin or in your body. This reaction occurs at random, so one bottle of shampoo might be loaded with nitrosamines and another bottle with the same ingredients sitting right next to it on the shelf may be absolutely safe. And you could take that safe bottle and use it with no ill effects, but it might create nitrosamines when used by another family member. Because nitrosamines are easily absorbed through the skin, you may get a higher exposure to this carcinogen from washing your hair than you would eating the infamous nitrite-cured bacon.

WHAT YOU CAN DO

If this were an advertisement for a new dandruff shampoo, you would read, "At last! You can get rid of dandruff forever!" I personally have never had dandruff, and so have never tried this method, but several of my friends swear by it. This sounds strange, I know, but it works:

- Instead of using shampoo of any kind, use baking soda (just baking soda and nothing else) to wash your hair. Just take a handful of dry soda and rub it vigorously into your wet hair, using the tips of your fingers to massage it into your scalp. Rinse thoroughly and dry hair. Stop using all shampoo, conditioner, hair spray, or any other product on your hair. Wash your hair at the same intervals you normally do, but keep using baking soda and no chemicals on your hair whatsoever. The first time you do this your hair might look like straw (don't try it right before a hot date) but just stick with it. After a few weeks, old dandruff scales will be gone, your scalp will begin to generate its natural oils, and your hair will get very soft. One of my friends said this made her hair feel better than any conditioner ever did. Once you get to this point, you can use a natural shampoo alternately with the baking soda. You will find your own balance of products to keep your hair beautiful and dandruff-free.

To find a natural shampoo, don't look at the drugstore; go instead to a natural food store. They are stacked to the ceiling with them. At last count, I could obtain locally 29 different brands that were either all-natural, or contained an acceptable minimum of relatively safe petrochemical derivatives. And this doesn't even include the number of natural shampoos available by mail. So there's no excuse for using unhealthful products on your hair.

You can also use any liquid soap for shampoo, or rub bar soap into your hair. You will need to rinse thoroughly, and if it leaves a residue, try a rinse with diluted vinegar or lemon juice. I have heard soap works well for dry or normal hair, but I don't find it satisfactory on my oily hair. My hair just doesn't get fluffy enough.

As long as we're discussing hair, here's a tip about hairbrushes and combs. If you have a problem with static electricity in your hair, it is probably due to the **nylon** bristles on your brush or your plastic comb. When you switch to a natural bristle brush and a wooden comb (both available at natural food stores and by mail), the problem will go away immediately.

FLUORIDE MOUTHWASH AND TOOTHPASTE ⊖ ☞

From the Manufacturers' Warning Labels:

> **CAUTION: Keep out of reach of children. For rinsing only. Do not swallow. Do not administer to any child under 6 years of age.**

Let's start with mouthwash first, since that's the product with the warning label.

Mouthwashes contain a number of ingredients that could be harmful or fatal if swallowed. I find it interesting that the same germ killers (**phenol, cresol,** and **ethanol**) that are used in bathroom disinfectants are also used (although in a lower concentration) in a product designed for use in the mouth, which could be swallowed. Ethanol is a petrochemical version of the same ethyl alcohol you drink in alcoholic beverages and can affect the central nervous system and cause nausea and drowsiness. If too much is taken, the body can go into shock or a coma, possibly resulting in death. Mouthwashes might also contain **formaldehyde, artificial colors, ammonia,** and **hydrogen peroxide.**

Mouthwash is one of those products it's easy to make assumptions about. You would think that if a product is intended for use in the mouth, it would also be safe to swallow. I was surprised to see the label warn against this.

Fluoride toothpastes have no warning labels, but may contain **ammonia, ethanol, artificial colors and flavors, formaldehyde, mineral oil,** sugar, and carcinogenic **polyvinylpyrrolidone plastic (PVP),** the same plastic resin used in hair spray.

Fluoride mouthwashes and toothpastes are often used by children as added protection against tooth decay. While there is no question that the optimal dose of fluoride will help prevent cavities, there is a possible danger here that with the fluoride in mouthwash and toothpaste and tap water, children might be getting *too much* fluoride, causing mottling of the teeth and many common ills.

Many mouthwashes and toothpastes, both fluoridated and unfluoridated, are sweetened with **saccharin,** which is known to cause cancer in laboratory animals. Food products and chewing gum containing saccharin require warning labels to this effect, but mouthwash and toothpaste do not.

Nonfluoride mouthwashes and toothpastes formulated to prevent bad breath contain many similar ingredients and pose the same type of hazards.

WHAT YOU CAN DO

Check with your dentist to see if your child needs fluoride treatment. It may be safer and more effective to have your dentist give periodic fluoride treatments in the office where the dose can be regulated, than to apply fluoride haphazardly with mouthwash and toothpaste.

I know of at least one natural toothpaste that contains the natural mineral calcium fluoride. It is widely available at natural food stores and by mail, as are other natural toothpastes. If you're not interested in fluoride in your toothpaste, you can brush your teeth with plain baking soda (or baking soda flavored with a few drops of your favorite extract or essential oil), or mashed strawberries. I know this sounds funny, but I have heard good reports about mashed strawberries. They foam up just like toothpaste and seem to contain an ingredient that helps people with gum problems. Brushing without toothpaste works fine too. The point is to get the food out from between your teeth.

Good oral hygiene, including regular brushing and flossing, will prevent bad breath, but if you are one of those people who likes to use mouthwash just because it makes your mouth taste good, you can get the same effect by rinsing your mouth with very strong, cooled mint tea (or any other flavor, for that matter), or with your favorite flavored extract mixed into pure water. You can also buy mouthwash made from natural ingredients at your natural food store or by mail. You should have no trouble finding one, as there are several popular brands on the market.

If you have a persistent problem with bad breath, consult your dentist.

It seems appropriate to make a comment here about toothbrushes. Most toothbrushes are made with **nylon** bristles. If you have gum problems, you might try switching to a natural bristle brush, available at natural food stores and by mail. These bristles are not only less abrasive, but some people with bleeding gums seem to react to the nylon itself, and their problem disappears when they switch tooth- brushes.

ASTRINGENTS

From the Manufacturers' Warning Labels:

> **CAUTION: For external use only. Avoid contact with eyes. In case of accidental ingestion, seek professional assistance or contact a poison con- trol center immediately.**

Most astringents are just **alcohol** (denatured to make it un- drinkable) with a little **artificial fragrance** and **artificial color** thrown in to make it more interesting, and a bit of glycerin to soften skin. Some astringents are very harsh and can burn skin, and the perfumes can irritate the skin and cause allergic rashes. Accidentally splashed in the eye, astringents can damage delicate membranes. Fumes from the alcohol can irritate nasal passages and cause nausea, drowsiness, and dizziness. If accidentally swallowed, astringents can be deadly, especially to a child.

WHAT YOU CAN DO

As a person with oily skin, I have been using astringents all my life. I remember using a very popular brand once that got rid of all the oil all right, but also made my face red and blotchy, and it burned so much I couldn't sleep all night.

Now, I only occasionally use a lovely lemon astringent I order by mail. It gets my skin clean and leaves it soft without harsh chemi- cals. My skin has been less oily in the last six years since I have been using a natural mint soap instead of a detergent soap bar, which strips the skin and makes it generate more oil.

I also occasionally use a clay mask to absorb excess oil. They are a luxurious treat to pamper yourself with. Go to the natural food store and buy pure clay; don't use a department store mask, which contains many chemicals. Mix clay with water to form a thin paste, apply, and relax until the clay dries. Regular applications of clay masks will reduce the overall oiliness of your skin, and will probably eliminate the need for an astringent.

You can probably find astringents made from natural substances at your local natural food store. There are several widely distributed brands available.

Because the active ingredient in most astringents is alcohol, you can use a cotton ball dampened with vodka for the same effect. Strong chamomile or mint tea (cooled) also have natural astringent properties and have a pleasant aroma. Using buttermilk will not only remove oil, but will leave your skin very soft. Simply apply with a cotton ball, wait 10 minutes, then rinse.

You can also make your own astringents at home with the following formulas. I know they sound funny, but try them and see how well your skin responds. Just process the ingredients as directed, then pour them into a glass spray bottle, and store under refrigeration.

- *Boil dark green lettuce leaves for 10 minutes in enough water to cover. Let cool and strain.*

- *Mix 1 part vodka with 9 parts strong chamomile or mint tea.*

- *Combine ⅔ cup pure water, 2 tablespoons vodka, and ¾ cup borax in a blender until borax is dissolved.*

- *Blend ¼ cup lemon juice, ¼ cup lime juice, ¼ cup pure water, and ⅛ cup vodka. Strain to remove pulp. This works especially well for very oily skin.*

NAIL POLISH AND NAIL-POLISH REMOVER

From the Manufacturers' Warning Labels:

> **CAUTION: Keep away from small children. Harmful if taken internally. In case of accidental**

ingestion, consult a physician or poison control center. Harmful to synthetic fabrics, wood finishes, and plastics.

Since nail-polish remover is the product that shows a warning label, let's start with that. The primary ingredient in most nail-polish removers is the solvent **acetone**, which can not only dissolve nail polish, but can cause your nails to become brittle and split, and skin rashes to develop on your fingers. When inhaled, the fumes from nail-polish remover can irritate your lungs and make you feel lightheaded. When accidentally ingested, acetone can cause restlessness, vomiting, and, ultimately, collapse into unconsciousness.

Nail polish is even more toxic, but, amazingly, doesn't have a warning label. Nail polish contains **phenol, toluene,** and **xylene**, three highly volatile and harmful chemicals, but its basic ingredient is a **formaldehyde** resin, which can cause discoloration and bleeding under the nail. My great-aunt used to own a drugstore, and when I was a teenager wearing nail polish she used to tell me stories of the women who would come into the store with their nails cracked and bleeding from wearing nail polish all the time. She said "their nails couldn't breathe," and warned me to remove the nail polish regularly to expose my nails to air.

WHAT YOU CAN DO

I haven't any suggestions for making nontoxic nail polish, and as of this writing I don't know of any brands that are any more natural or less toxic than others.

If you want to wear nail polish, the most dangerous part of your exposure is during application. Once it is dry, the only hazard is to your nail and surrounding skin. Apply nail polish in a well-ventilated area, preferably outdoors.

Because no natural substance can melt nail polish, if you are going to wear polish, you have to use nail-polish remover. Use only in a well-ventilated area, and wash your hands thoroughly immediately after use to remove any traces that may irritate skin.

HAIR-REMOVAL PRODUCTS ☞ 👁 ✋ ℄

From the Manufacturers' Warning Labels:

> **CAUTION:** Do not apply near eyes, on inflamed, chapped, broken or newly tweezed areas, or the vaginal area. If cream gets in eyes, flush thoroughly with lukewarm water.

The most dangerous ingredient in hair-removal products is **ammonium thioglycolate,** which can cause severe skin rashes, swelling, redness, and the breaking of small blood vessels under the skin.

WHAT YOU CAN DO

Instead of using a chemical product, remove unwanted hair by shaving, tweezing, or electrolysis.

You can remove facial hair with natural beeswax. Buy it in chunks at a natural food store or hardware store, or an uncolored beeswax candle will work fine too. Just melt a small amount in a pan until it's very warm but still cool enough to touch. Dust your skin with body powder and apply the warm wax with a wooden spatula. Allow the wax to dry for a few seconds, then remove it quickly with a light tapping. The hairs will come out with the wax. It hurts about as much as when you pull a band-aid off your arm. Soothe your skin with a bit of unscented cream or lotion.

HAIR COLOR ☞ 👁 ✋ ℄

From the Manufacturers' Warning Labels:

> **CAUTION:** This product contains ingredients that may cause skin irritation on certain individuals, and a preliminary test according to accompanying directions should first be made. This product must not be used for dyeing eye-

lashes or eyebrows: to do so may cause blindness.

WARNING: Contains an ingredient that can penetrate your skin and has been determined to cause cancer in laboratory animals.

If the FDA had its way, this warning would appear on all packages of hair coloring. Manufacturers, however, are able to get around the warning labels by just slightly reformulating their products to remove the carcinogen, but then they replace it, quite legally, with another chemical that is just as dangerous. It takes four years or more to test the new chemical for safety before the FDA can propose another warning label, and many manufacturers take full advantage of the lag time.

Regardless of the toxicity of the chemicals used, hazardous hair-coloring products cannot be banned from the marketplace because of a 1938 law, still in effect, which was passed when the hair dye industry persuaded Congress to exempt these products from government regulation. At that time, the synthetic coal-tar dyes used were known to cause serious allergic reactions in some users. Because government regulation of these hazardous products would have been a major threat to the entire hair dye industry, manufacturers lobbied relentlessly until they won the exemption they needed to stay in business. And whatever it's taken to keep that law on the books, they've done it!

An investigation done by *Consumer Reports* magazine revealed about *twenty* different chemicals used regularly in hair-coloring products that are potential human carcinogens. Hair dye products may also contain **coal-tar dyes, ammonia, detergents, hydrogen peroxide,** and **lead.** The scalp is very porous, and these chemicals can be easily absorbed.

WHAT YOU CAN DO

The safest commercial hair color is henna, a powder made from a plant source (available at most pharmacies and natural food stores). Many different shades are available to darken or highlight your hair. Henna gives hair a semipermanent protein coating, which washes out gradually over a six-month period. There are a few hair-coloring products made from natural ingredients. Check with your natural-food store.

You can color your hair temporarily with natural rinses made from plant materials. Women used these rinses for centuries before chemical dyes were invented. These natural products will not change your hair color as drastically as chemical formulas, but they can highlight and enrich your hair color in beautiful and subtle ways that no chemical dye can duplicate. When the rinse is used repeatedly, the color will deepen in intensity each time. Once your hair is the shade you like, use the rinse periodically to keep color from fading.

To use the following formulas for plant-based hair rinses, process ingredients as directed, then strain and cool before using. Pour the liquid through the hair 15 times, catching it in a basin below and rerinsing with the same liquid. Wring out any excess and leave in the hair for 15 minutes before a final rinse with clear water.

Blonde

Note: The following rinses will lighten the hair to an even greater degree if you dry your hair in the sun.

- Mix 1 tablespoon of lemon juice in 1 gallon of warm water.
- Or, simmer 4 tablespoons of chopped rhubarb root in 3 cups of hot water for 15 minutes.
- Or, steep ½ cup yellow-blossomed flower or herb (try chamomile, calendula, mullein blooms and leaves, yellow broom, saffron, turmeric, or quassia chips) in 1 quart boiling water for ½ hour. (You can also take this rinse and mix it with equal parts of lemon juice, add a little arrowroot, and stir over low heat until it forms a gel. Apply the gel to your hair and sit out in the sun for an hour.

Brown/Brunette

- Rinse hair with a strong black tea or black coffee.
- Or, cook an unpeeled potato, apply the cooking water to your hair with a cotton ball, keeping the rinse away from your skin to prevent discoloration.

To Cover Gray

- Simmer ½ cup of dried sage in 2 cups water for 30 minutes, then steep for several hours. Apply the tea to hair and leave it on until your hair dries; then rinse and dry your hair again. Apply weekly until desired shade is achieved, then monthly to maintain color.

- Or, cover crushed black walnut shells with boiling water, add a pinch of salt, and soak for three days. Add 3 cups boiling water and simmer in a glass pot for 5 hours, replacing water as needed. Strain and simmer the liquid until it has reduced to one-quarter the original volume. Add 1 teaspoon ground cloves or allspice and steep in the refrigerator for one week, shaking the jar every day. Strain and use carefully, as this mixture will stain everything you touch. Wear gloves and try to avoid skin contact.

Red

- Make a strong tea of rosehips or cloves, or use strong black coffee.

- Or, steep together in boiling water for 15 minutes 1 tablespoon each of henna, chamomile flowers, and vinegar.

If you absolutely must use a commercial hair dye, *Consumer Reports* magazine suggests you take the following precautions:

- Don't use hair dyes more often than necessary. Once every four to six weeks should be frequently enough.

- Don't leave the dye on your head any longer than necessary according to package instructions.

- Flood your scalp thoroughly with water after applying dye.

- Use a technique that involves minimum contact between the dye and your scalp (the dye is absorbed into the bloodstream very easily through hair follicles and oil glands).

- Put off using any hair dyes to as late in life as possible.

PERMANENT WAVES

From the Manufacturers' Warning Labels:

> CAUTION: Keep out of reach of children. Avoid getting waving lotion in eyes. If it does, rinse eyes with water. In case of accidental ingestion, consult a physician immediately.

Ammonium thioglycolate is the most hazardous ingredient in permanent-wave solutions. It can cause skin rashes on your hands and scalp, redness, swelling, and hemorrhages under your skin. Because the scalp is very porous, this substance can easily be carried into the bloodstream.

In addition to the skin effects, permanent waves smell very strongly of ammonia, which can cause breathing difficulties and coughing.

WHAT YOU CAN DO

If you want to perm your hair, there are several ammonia-free home perms available made with natural ingredients. Your natural food store may carry them; if not, they can be ordered by mail.

SUPERABSORBENT TAMPONS

From the Manufacturers' Warning Labels:

> ATTENTION: Tampons are associated with Toxic Shock Syndrome (TSS). TSS is a rare but serious disease that may cause death.

The symptoms of Toxic Shock Syndrome (TSS) include a fever of 102° F or more, vomiting, diarrhea, a sunburn-type rash and subsequent skin peeling, and a rapid drop in blood pressure that may cause fatal shock. Women who have already had TSS or have recently delivered a baby should not use tampons at all, since they are at greatest risk.

The exact cause of TSS is unknown, but it has been determined that staphylococcus bacteria are present in most cases. Although TSS has been associated with all tampon use, the brands made from super-absorbent fibers seem to pose the highest risk.

WHAT YOU CAN DO

Choose a brand of tampon that is made without superabsor-bent fibers. Materials used in tampon construction are listed on the label, so look for brands that are made of cotton or rayon. Some labels even state "no superabsorbent fibers."

A study published in the *American Journal of Obstetrics and Gynecology* reported that the presence of superabsorbent fibers significantly increased the rate of growth of staphylococcus bacteria, and that production of the bacteria dramatically decreased when it was grown on cotton fibers.

Small, soft sea sponges can be inserted as a "natural" tampon. Federal regulations prohibit the marketing of sponges "to insert into the body," so you can't go into a store and ask for "tampon sponges." What you want to look for are "cosmetic sponges" which look like fluffy pillows and are used to apply foundation makeup.

Before using a sea sponge as a tampon, rinse it first several times in running water, then sterilize it by boiling it for 2 minutes in a pot of plain water. Wash your hands before inserting or removing the sponge. After removing it, rinse the sponge well in running water and squeeze it dry. It can then be reinserted. Frequent removal and rinsing is highly recommended. If you have trouble removing your sponge, tie a piece of cotton quilting thread around the sponge to help pull it out.

Sponges should be stored in a clean, airy location when not in use. Do not put a damp sponge in a plastic or airtight container. Resterilize your sponge before each period to prevent bacteria growth.

I have never tried sea sponges for this purpose (I am content with natural fiber tampons), but I have several friends who think they are wonderful. They say they are very comfortable with them and . . . they're never caught without a tampon.

Of course, you could also use feminine napkins, but if you do, stay away from the deodorant variety. More than 20 percent of the respondents in a survey done by *Consumer Reports* magazine indicated that they had been warned by their doctors not to use deodorant tampons or pads. Even people who are not normally allergic to perfume can, over time, develop irritations to these deodorant scents. Your natural food store may carry disposable or reusable menstrual pads made from 100 percent cotton, or you can order them by mail (see Resources).

CONTRACEPTIVES

From the Manufacturers' Warning Labels:

> Cigarette smoking increases the risk of serious cardiovascular side effects from oral contraceptive use. This risk increases with age and with heavy smoking (15 or more cigarettes per day) and is quite marked in women over 35 years of age. Women who use oral contraceptives should be strongly advised not to smoke.
>
> The use of oral contraceptives is associated with increased risk of several serious conditions including thromboembolism, stroke, myocardial infarction, liver tumor, gall bladder disease, visual disturbances, fetal abnormalities, and hypertension.

Oral Contraceptive Pills

Oral contraceptive pills contain synthetic female hormones that prevent pregnancy by altering the chemical balance in your body. The hormones trick the body into thinking it is pregnant, causing the release of eggs to be suppressed.

Many women are uncomfortable about taking a hormone-changing medication day in and day out for years on end, and they should be. The list of side effects is lengthy: weight gain, acne, eczema, nausea, gum inflammation, inflammation of the optic nerve (leading to loss of vision, double vision, eye pains and swelling, and an inability to wear contact lenses), headaches, vaginal yeast infections, depression, loss of sex drive, formation of blood clots, heart attacks, high blood pressure, stroke, gall bladder disease, liver tumors, birth defects, ectopic pregnancy, skin cancer and cancer of the reproductive organs, breast tumors, menstrual irregularities, post-pill infertility, and infections.

According to the FDA, the Pill should not be used by women who have or have had blood clots in their legs or lungs, have pains in the heart or have had a heart attack or stroke, know or suspect they have cancer of the breast or sex organs, have unusual vaginal bleeding that has not yet been diagnosed, or know or suspect they are pregnant.

Intrauterine Devices

As of this writing, there is only one intrauterine device (IUD) on the market. Others have been withdrawn from sale because of negative health effects or because the high cost of defending lawsuits made their sale economically unfeasible for the manufacturer.

The most serious complication associated with IUD use is permanent sterility. The risk is so high that an article in the *New England Journal of Medicine* recommends that a woman who has never borne a child use the IUD as a contraceptive only if she does not wish to ever conceive, and that IUD use is appropriate for women who have borne children if they do not want to have more children and do not wish to undergo sterilization.

Spermaticides, Diaphragms, and Contraceptive Sponges

The diaphragm and vaginal spermaticides (including foam, creams, jellies, suppositories, and foaming tablets) all pose the same hazards, because a diaphragm must be used with a chemical spermicide. Few side effects beyond local irritation have been reported by users of these products, but their use is being connected with cases involving birth defects and spontaneous abortion. In one court case, a mother and child were awarded $5.1 million when the judge ruled that a popular contraceptive jelly caused the child's birth defects, and that the product should have carried a warning label stating that babies conceived while using spermicide may have birth defects. Spermaticides do not always kill sperm, and, sometimes, damaged sperm go on to fertilize eggs, resulting in deformed babies.

United States congressional hearings found evidence to suggest that spermaticides might be carcinogenic and mutagenic.

An FDA advisory review panel studying the safety of vaginal spermaticides found only a few out-of-date studies that evaluated the safety of spermaticides. None of these studies addressed the issues of the effects of spermaticides on an unborn fetus, or possible genetic mutation, carcinogenicity, or toxic effects.

Because manufacturers are only required to list active ingredients on the label, it is hard to tell exactly what might be in a spermaticide product. Additional ingredients could include **alcohol, formaldehyde, methylbenzethonium chloride, perfume,** and **preservatives,** among others.

Contraceptive sponges contain an additional health hazard. Studies done by the World Health Organization (WHO) showed a high frequency of cancer in mice given daily insertions of a polyurethane sponge, the same material with which contraceptive sponges are made. Although after much testing the FDA approved the sponge in June 1983, they almost immediately required package labels to carrying this warning regarding Toxic Shock Syndrome:

Warnings:

- If you experience two or more of the warning signs of toxic shock syndrome (TSS), including fever, vomiting, diarrhea, muscular pain, dizziness, and rash similar to sunburn, consult your physician or clinic immediately.

- The sponge should be removed within the specified time limit.

- In case of accidental ingestion of the sponge, call a poison control center, emergency medical facility, or doctor.

The hazards of chemical spermaticides are especially important to consider in connection with use of the sponge, as it contains 10 times more spermicide than is used with a diaphragm.

WHAT YOU CAN DO

Condoms

Condoms are by far the safest commercial contraceptives—the only risk involved is a remote chance of allergic reaction to the latex rubber they are made of, or to the chemicals in the lubricants. They also have an added benefit, as they protect against sexually transmitted diseases (including herpes and AIDS), as well as preventing pregnancy.

The biggest objection expressed about using condoms is that, obviously, the sexual experience is not quite the same as it is without them. A male friend told me once it was like taking a shower with a raincoat on. The way around this is for both partners to cooperate, using them only when it's necessary. With natural birth control, you can have the best of both.

Natural Birth Control

Many couples today are choosing to use natural family planning (sometimes called the Fertility Awareness Method), which gives them freedom and protection without harmful side effects.

This method is based on calculating fertile times through monitoring vaginal mucus and changes in body temperature. After using the method, you may also notice subtle changes in mood, skin texture, and weight. Abstinence is required for all the days between your menstrual period and ovulation, approximately two weeks later.

This method can be very effective, but it requires diligent practice and a willingness to abstain. Some couples get frustrated waiting for the "safe days." It may work best for you to monitor body changes (it takes less than a minute each day once you know what to look for), have fun on the safe days, and use condoms on the questionable days leading up to ovulation. As an extra precaution against pregnancy, you can refrain for a couple of days during your fertile time to avoid accidents from faulty condoms.

I have purposely not explained the method fully here because you really should read at least two books on the subject and/or take a class before relying on natural birth control to be effective. Ask your gynecologist, a women's health clinic, or local family-planning center how you can get more information. You local library should have some books that give instructions.

PERFUME AND AFTERSHAVE

Perfumes and aftershave lotions are products that should be labeled "keep out of reach of children," but aren't. According to my local poison control center, because of the high alcohol content of these products, it takes only about a tablespoon ingested accidentally for a small child to become intoxicated. This leads to lowered blood sugar, which can cause unconsciousness, and eventually the child could fall into a coma and die, all from one tablespoon of your favorite fragrance.

Although there has been very little scientific study on the health effects of wearing scented products, they are generally recognized as being highly allergenic, and are notorious for causing mild to severe skin irritation. Fragrances are also a common cause of headaches and nausea. Almost all of us have been in a situation where we have been near someone who was wearing just too much perfume and we felt sick to our stomachs.

Labels on perfumes and aftershaves are somewhat incomplete, because many chemicals are used during the manufacture of perfume that do not end up on the label. In addition to **artificial fragrance** and **alcohol**, scented products may contain **formaldehyde, phenol, trichloroethylene,** and **cresol,** all toxic solvents.

WHAT YOU CAN DO

For many people, fragrance is an integral part of personal expression. It *is* possible to use pleasant *nontoxic* scents.

Look for essential oils at your natural food store or order them by mail. When buying essential oils, be sure to find out if the oil has been derived from a natural source. A number of confusing terms are used to describe essential oils, so let's sort them out.

Natural essential oils are derived from plant sources, and the resulting fragrance is that of the plant. A *natural* lemon scent, for example, would come from a part of the lemon plant. *Synthetic* oils are also from plant sources, but the fragrance is not that of the original plant. A *synthetic* lemon scent might not be from a lemon at all, but from a geranium! *Artificial* oils, also called *perfume oils,* are made from petrochemicals.

Essential oils are also labeled according to their dilution. Those labeled *true, absolute,* or *concrete* are pure oils; *extracts* or *tinctures* have been diluted with grain alcohol; and *extended oils* contain plasticizers.

Essential oils are extremely potent and must be used with care. Because they are so powerful they are very irritating to the skin and should not be applied without dilution. Do not apply to mucous membrane areas at all. Before using, dilute essential oil by adding a few drops to one ounce of vegetable oil or vodka. As with any other perfume product, keep these out of children's reach.

Many people who think they are allergic to perfume find that they can tolerate natural fragrances very well. I am allergic to flower fragrances, no matter what the source, but I have a lot of fun with citrus and herbal scents. I have my shampoo custom-scented with French vanilla and spearmint, an unusual but wonderful combination that really wakes me up in the morning!

ANTIPERSPIRANTS AND DEODORANTS

Antiperspirants may contain **aerosol propellants, ammonia, alcohol, formaldehyde**, and **fragrance**, but the primary danger is the active ingredient that helps stop wetness: **aluminum chlorohydrate.**

Aluminum chlorohydrate can cause infections in the hair follicles of the armpit and skin irritations that can be severe enough to require medical attention. There is also some question as to whether the aluminum salts in antiperspirants might contribute to a buildup of aluminum in the body. Aluminum from other sources has been associated with various brain disorders, including Alzheimer's disease. There is also some concern about the safety of using aluminum in an aerosol spray. Because aerosols produce airborne particles that are likely to be inhaled, there is a good chance that bits of aluminum will enter the lungs, and accumulate over time. Because the long-term health effects are unknown, aluminum-containing aerosol antiperspirants are only conditionally approved by the FDA for safety.

Nonantiperspirant deodorants may contain the bacteria-killing ingredient **triclosan**, which can cause liver damage when absorbed through the skin.

Antiperspirants and deodorants historically have a bad track record. In a recent ten-year period, more than eight different ingredients were banned by the FDA or voluntarily removed from products because they posed a threat to users.

WHAT YOU CAN DO

There are a number of natural deodorants on the market, but what works best is baking soda—just plain baking soda. I have been using it for 6 years and nobody has ever complained. In fact, I have recommended it to a number of people who have suffered for many years with unconquerable body odor, and they say it's the only thing that has ever worked for them. This works so well that I don't even want to suggest anything else. Just take a bit of dry baking soda on your fingertips and pat it under your arms when you've dried off after your shower. Your skin will be slightly damp, but not wet. If it feels too abrasive to you, you can mix it with cornstarch or white clay.

BUBBLE BATH

Bubble baths consist basically of just **detergent** and **artificial fragrance**, but the FDA receives many complaints regarding their use. Skin rashes, irritations, and urinary-tract, bladder, and kidney infections are commonly reported by users.

WHAT YOU CAN DO

Long, warm, relaxing baths are one of the most luxurious experiences. We should be healed and pampered by taking a bath, not harmed.

There are many lovely, natural things you can add to your bath. Here are some suggestions:

- ½ cup or more of baking soda (this is particularly soothing for a sunburn or skin rash)
- 1 cup Epsom salts
- 1 quart of whole or skim milk, or buttermilk
- slices and juice of several lemons or limes, or an orange or grapefruit
- 5 to 10 peppermint, chamomile, or other herb tea bags. Let them steep about five minutes in very hot water, then add to warm bath water.
- Any fragrant dried herbs (try lavender or rosemary) or your favorite sweet-smelling fresh flower petals. For an exotic bath, try ground or crushed spices (cloves, cardamom seeds, allspice, nutmeg, cinnamon, or ginger).
- Champagne, white wine, port, sherry, or other alcoholic beverages.
- And, if you want bubbles, here's what to do. Start running warm water into your tub and add ½ to 1 cup sodium hexametaphosphate (a natural mineral powder you might have already ordered from a chemical supply house to use as a dishwasher detergent). Sodium hexametaphosphate is the active ingredient used in a popular bath product—it makes the water feel wonderfully soft and slippery. The amount you will use depends on the hardness of your water, and the more you use, the more bubbles you will get. Swish the water around with your hand to dissolve the sodium hexametaphosphate and then start adding liquid soap, a few tablespoons at a time, right near the faucet where there is a lot of churning water. You will have lots of bubbles. For fragrance, you can use a naturally-scented soap, or add slices of lemon, a few drops of essential oil, dried or fresh herbs, or flower petals.

DEODORANT SOAPS ☺ ◉ ✑ ℔

Because deodorant soaps are considered to be over-the-counter drugs and not cosmetics, it's hard to tell exactly what ingredients might be in any one bar of deodorant soap. It might contain **ammonia, formaldehyde,** or **phenol,** and most probably contains **artificial fragrance.**

Deodorant soaps work by killing the bacteria that supposedly cause body odor. The most commonly used bactericide is **triclocarban.** While the FDA has not yet proved this chemical to be hazardous to the general public, they are concerned about the unknown effects of using deodorant soaps on a day-in, day-out basis, and there is an unresolved suspicion that triclocarban may cause cancer. The FDA also warns against using deodorant soaps on infants under 6 months of age.

Whether or not triclocarban is safe is not the only issue in question. Although triclocarban's effectiveness in killing bacteria has been well demonstrated, *there is no clear correlation between reduced bacteria counts and reduced body odor,* which makes me ask "Why use a deodorant soap at all?"

WHAT YOU CAN DO

Body odor is best prevented by regular bathing with plain soap and hot water. A bit of baking soda under your arms will neutralize any odors produced during the day.

Choose the purest soap you can find, preferably one that is unscented and uncolored. Some people feel soap is "drying to the skin," and for some soaps, especially those made from coconut oil, this is true. But there are so many mild soaps available made from natural glycerin, olive oil, or tallow, that you should have no problem finding one you like. You can find pure soaps at natural food stores, bath shops, drugstores, hardware stores, and cookware stores, or you can order a couple dozen different varieties by mail.

COSMETICS ⇔ ☞ 👃

Let's start with lipstick, because it's the most toxic. First, lipstick may contain several substances known to cause cancer in animals: **polyvinylpyrrolidone plastic (PVP), saccharin, mineral oil,** and **artificial colors.** These are not yet scientifically proven to cause cancer in humans, but it is known that almost all substances that cause cancer in humans also cause cancer in animals. So it's logical to assume there is a danger here. Are you surprised to see mineral oil on that list? There is enough evidence supporting its danger to make it forbidden as a food coating in Germany, even though its use in food products is still allowed in the United States. Lipsticks also frequently contain **artificial fragrance,** which can be very drying to the lips. Unlike other cosmetics, lipstick poses a double danger; in addition to absorbing it into their skin, women who wear lipstick often ingest a good deal of it as they speak, lick their lips, drink, and eat throughout the day. While I was unable to find studies on the health effects of lipsticks, I am concerned about the possible long-term effects of everyday exposure to these chemicals.

The next most dangerous cosmetic is mascara. It can contain **formaldehyde, alcohol,** and various **plastic resins.** Here, the primary danger is eye irritation, redness, burning, and swelling. We want to enhance our eyes, not hurt them.

The greatest danger from eyeshadow, powdered blush, and face powder is **talc.** Talc may be contaminated with carcinogenic asbestos, and as we apply these powders, they float through the air and end up in our lungs. Of course, many contain **artificial fragrance** also, which is a common cause of allergic reactions to cosmetics. Many liquid foundation makeups contain **mineral oil,** which is, as I mentioned before, a suspected human carcinogen.

WHAT YOU CAN DO

Yes, you can still wear makeup. There are a number of good brands of more natural cosmetics on the market, and as the demand increases, they look more and more like the department store brands.

If your natural food store doesn't carry them, you can order them by mail.

Most "natural" brands, however, will be colored with artificial colors, so read labels carefully. A couple of brands use natural colorings; they are difficult to find, but again, can be ordered by mail.

At the very least, use unscented "hypoallergenic" cosmetics.

My favorite cosmetics are made from colored clays, which are available at most natural food stores. Unless I am having my picture taken or have a television appearance, I don't generally wear foundation makeup. I just brush a little rose-colored cosmetic clay onto my cheeks with a big fluffy brush, put a little charcoal grey, blue, or brown clay around my eyes, brush on the least toxic mascara I can find, and put a little gloss on my lips that is tinted with natural carmine red. It gives me a finished look and enhances my natural coloring with other natural colors. Instead of lipstick, you could also get a clear gloss from your natural food store and apply it to your lips over a bit of cosmetic clay, or after you've "stained" your lips with beet or berry juice. Try it. I was surprised to see how beautiful the colors come out. I had been looking for a certain shade of reddish lipstick for years, and finally found it in my 100 percent natural lipgloss.

TALCUM POWDER

The danger associated with the use of talcum powder is simple —it may be contaminated with carcinogenic asbestos fibers. As you pat it on your body, you generally surround yourself with a good cloud of the stuff and a certain portion ends up in your lungs. There is *no* safe level for asbestos exposure.

HOW TO PROTECT YOURSELF

If you wish to use a body powder, there are a few available at your natural food store and by mail, but it might be just as easy, and a lot less expensive for you to make it at home.

The simplest thing is to just use plain cornstarch, rice starch, oat flour, arrowroot, or white clay, all available at your natural food store. You could also use eggshell powder, although this is a bit more trouble to make. You have to save your eggshells for a while, rinse and dry them thoroughly, and then crush them into a fine powder in a blender or food processor.

You can also add ground herbs to your body powder for a fresh and natural fragrance.

DISPOSABLE DIAPERS

The primary danger from disposable diapers is that they contribute to diaper rash. They are made from **synthetic fibers** and other **plastics** and unknown deodorizing chemicals that are not compatible with a newborn baby's skin. In addition to diaper rash, complaints made to the Federal Consumer Protection Agency include chemical burns, noxious chemical and insecticide odors, reports of babies pulling the diapers apart and putting pieces of plastic into their mouths and noses, of plastic covers melting on baby's skin, and the chemical dyes on the diapers staining the baby's skin.

WHAT YOU CAN DO

Use 100 percent cotton cloth diapers with natural-fiber diaper covers. Even if you send the diapers out to a diaper service to be washed, you wind up spending less than you would buying disposables, and your baby will be so much more comfortable.

If you wash the diapers yourself, use the hottest water in your washing machine and wash with a gentle soap and borax (in the laundry supplies department at your supermarket). Detergent residues can contribute to diaper rash, too. Then run the diapers through the hottest setting on your dryer for 40 minutes or longer. This should be sufficient to kill bacteria. Whenever, possible, air the diapers outdoors in sunlight, which acts as a natural disinfectant. Soaking soiled diapers in a mixture of ½ cup borax added to a pail of warm water will reduce odors and staining, and will make diapers more absorbent.

TOILET PAPER AND TISSUES

There are three main hazardous ingredients in toilet paper and tissues: **formaldehyde, artificial fragrance,** and **dyes.** I have not seen any formal studies on the health effects of toilet paper, but several years ago, "Dear Abby" ran a number of letters describing complaints.

One woman went to her family doctor with symptoms and was sent to a gynecologist, who told her she had genital herpes. For 10 years she had periodic flareups, typically after her periods or after sexual activity. She described the emotional strain and frustration this experience caused her. As a last resort, she saw another gynecologist. This doctor accurately diagnosed that she was allergic to the dye in her colored toilet paper. She's been happily symptom-free ever since.

Another reader wrote that for a year the itching and burning were unbearable. She was told she had some kind of infection, but no one would specify what kind. Finally, a nurse suggested the problem might be her colored toilet paper. One week after changing to white toilet paper the symptoms disappeared.

Yet another letter was written by a man who stated that shortly after his wife switched to a scented toilet paper, he and his young daughter experienced pain in the genital area. Luckily, they put two and two together and have had no problems since returning to unscented toilet paper.

WHAT YOU CAN DO

The best you can do with toilet paper is to buy it plain white and unscented. It will still contain formaldehyde, as all wet-strength papers do, but I can't think of any substitute.

For tissues, again buy them plain white and unscented, or use cotton handkerchiefs, which are inexpensive and easy to find.

CHAPTER 7

Food ☙

Finally, the chapter on food. Are you surprised there are so many other household items that are more harmful? With all the emphasis on pesticides and chemical contaminants in foods today, you would think that the food we eat is the source of the most dangerous exposure we have to toxics. Although I don't want to diminish the importance of hazardous substances in foods, in relation to other types of products, foods are so well regulated that the immediate risks are fairly low, but the long-term dangers must be considered. With careful food selection, storage, and preparation, we *can* have a healthful food supply.

The optimal diet for minimizing your exposure to toxic substances in foods is based on fresh, whole foods including organically grown fresh fruits and vegetables, organically grown whole grains and legumes, eggs from free-range hens, lean meat or poultry from free-range animals, nonfat milk and low-fat cheeses, ocean fish, and honey, date sugar, barley malt syrup, fig syrup, and rice syrup (in small quantities). You can buy all of these foods at a large natural food "supermarket," or by mail.

In their *Dietary Guidelines for Americans,* the U.S. Department of Agriculture and U.S. Department of Health and Human Services suggest a diet very similar to the one outlined here. These guidelines were developed to reflect new scientific evidence that a good diet can promote health and help prevent common diseases such as cancer, heart disease, and stroke. I'm not surprised that this "good health" diet is also relatively low in chemicals. Here's what they recommend:

1. Eat a variety of foods daily, including selections of
   - Fruits
   - Vegetables
   - Whole grain and enriched breads, cereals, and grain products
   - Milk, cheese, and yogurt
   - Meats, poultry, fish, and eggs
   - Legumes (dry peas and beans)
2. Maintain ideal weight
3. Avoid too much fat, saturated fat, and cholesterol
   - Choose lean meat, fish, poultry, dry beans, and peas as your protein sources
   - Moderate your use of eggs and organ meats (such as liver)
   - Limit your intake of butter, cream, hydrogenated margarines, shortenings and coconut oil, and foods made from such products
   - Trim excess fat off meats
   - Broil, bake, or boil rather than fry
   - Read labels carefully to determine both amount and types of fat contained in foods
4. Eat foods with adequate starch and fiber
   - Substitute starches for fats and sugars
   - Select foods that are good sources of fiber and starch, such as whole-grain breads and cereals, fruits and vegetables, beans, peas, and nuts
5. Avoid too much sugar
   - Use less of all sugars, including white sugar, brown sugar, raw sugar, honey, and syrups

- Eat less of foods containing these sugars, such as candy, soft drinks, ice cream, cakes, and cookies

- Select fresh fruits or fruits canned without sugar or light syrup rather than heavy syrup

- Read food labels for clues on sugar content—if the names sucrose, glucose, maltose, dextrose, lactose, fructose, or syrups appear first, then there is a large amount of sugar

- Remember, how often you eat sugar is as important as how much sugar you eat

6. Avoid too much sodium

- Learn to enjoy the unsalted flavors of foods

- Cook with only small amounts of added salt

- Add little or no salt to food at the table

- Limit your intake of salty foods, such as potato chips, pretzels, salted nuts and popcorn, condiments (soy sauce, steak sauce, garlic salt), cheese, pickled foods, cured meats

- Read food labels carefully to determine the amounts of sodium in processed foods and snack items

7. If you drink alcohol, do so in moderation

Changing your eating habits needn't be drastic—you don't have to do it overnight. The idea is to start choosing fresh, whole foods more often, and choose packaged foods less often. You might start simply by eating more fresh fruits and vegetables each day, or by drinking low-fat or nonfat milk instead of whole milk. Substitute fresh fruit for desserts. Buy whole-grain breads instead of white bread.

Food patterns are hard to change because we have so many emotions tied to them. We remember our favorite dishes from childhood; we want to socialize with friends. We have a basic instinct to eat certain foods, and it's difficult to stop and think about the potential hazards of the food we are putting in our mouths. So just take your time. Even if you just read a few labels next time you go to the store, and maybe change your brand of pickles, it's a good place to begin.

Creating a healthful diet would be simple if this were all we needed to know about food—but it isn't. We need to learn how to choose additive-free packaged foods for those occasions when we just don't have time to cook, and we need to know what's in our food that's not on the label. For all the label-reading we are encouraged to do, the most dangerous substances found in our food supply are not even listed there. They are used in the manufacture of our food, or enter accidentally from contaminated soil, air, and water. We need to know these risks, and what to do about them.

PACKAGED FOODS ☜

The easiest place to begin removing harmful substances from your diet is with packaged foods that contain food additives. Because manufacturers are required by law to list all food additives on package labels, if they are used in a food product as an ingredient, they are simple to spot and thus avoid.

Sulfites pose the most immediate danger of all additives. Used primarily to reduce or prevent spoilage and discoloration, they can trigger severe allergic reactions in sulfite-sensitive individuals. These reactions include breathing difficulties, wheezing, vomiting, nausea, diarrhea, unconsciousness, abdominal pain, cramps, and hives. In some cases, anaphalactic shock can result in immediate death. Asthmatics are the primary group at risk, and it is estimated that about 10 percent of all asthmatics are sulfite-sensitive. However, one-fourth of the complaints received by the FDA involved individuals who have no history of asthma. Since the FDA considers this additive to be safe for consumption by the general public, it is not banning its use, and you must take the responsibility for controlling your intake.

Sulfites appear on package labels as **sulfur dioxide, sodium sulfite, sodium and potassium bisulfite,** and **sodium and potassium metabisulfite.** They are most commonly found in dried fruits, shellfish (fresh, frozen, canned, or dried), soups, wine vinegar, vegetables (fresh, peeled, frozen, canned, or dried), packaged lemon juice, avocado dip, maraschino cherries, packaged fresh mushrooms, canned foods, fruit juices, gelatin, potatoes (fresh, peeled, frozen, dried, or canned), salad dressings, sauces and gravies, and corn syrup.

Sulfites are almost always used in wine and beer production, and are *not* listed on the labels of alcoholic beverages. The most common exposure to sulfites, however, is in restaurants, where the kitchen may use them to keep foods looking fresh when they're really not. Sulfites are frequently used in salad bars, on fresh fruits and vegetables, precut potatoes, seafood, cooked vegetable dishes, and bakery products *without informing the customer.*

Nitrates are the next most dangerous because they can cause cancer. Nitrite is added to 60 to 65 percent of all pork produced in the United States, as well as some other meat, poultry, fish, and cheese. It is especially prevalent in processed meats such as bacon, sausage, luncheon meats, and hot dogs to preserve the pink color and inhibit the growth of the bacteria that causes botulism food poisoning. Nitrates themselves are not particularly harmful, but when they combine with amines in the food, they form carcinogenic **nitrosamines.**

Artificial colors also pose a cancer risk—almost all have been shown to be carcinogenic in animal studies. According to the National Cancer Institute, "We should assume that agents that cause cancer in animals are likely to cause cancer in humans. To prevent cancer, we cannot afford to wait for absolute proof of carcinogenicity in humans. Instead, we must heed the warnings provided by laboratory animal experiments and reduce or eliminate human exposure to probable cancer-causing agents."

But cancer, though certainly a major concern, is not the only problem. For, along with artificial colors, **BHA, BHT,** and **artificial flavors** have been linked to hyperactivity and behavioral disturbances in children by Dr. Benjamin Feingold of Kaiser-Permanente Medical Center. While it has been very difficult to substantiate this with scientific studies, the observations of parents and doctors over the past fifteen years confirm that avoiding artificial additives has significantly improved their children's conditions, making this evidence hard to ignore.

MSG is a popular additive that gained widespread use as a flavor enhancer before its healthfulness came into question. However, many users probably wondered privately about the safety of this additive, since its most well-known side effect is Chinese Restaurant Syndrome—a condition characterized by numbness, weakness, heart palpitations, cold sweat, and headache. Not everyone has the reactions and not everyone experiences symptoms of the same intensity, but I have heard from several friends that have been to China that when you eat MSG three times a day, every day, you really start to feel it! In addition, animal studies show that MSG can cause brain damage, stunted skeletal development, obesity, and female sterility. It is on the FDA list of additives that need further study for mutagenic and reproductive effects.

Also on the EPA list for further study is the preservative **ethylenediamine tetraacetic acid (EDTA)**. Toxicology books list many possible symptoms associated with large exposures including dizziness, sneezing, headaches, nausea, and asthma attacks. The effects caused by the amounts ingested in food are still being investigated.

Most processed foods contain **sugar** or **salt**. Although moderate amounts of either of these substances are not particularly harmful for most people, the amounts of sugar and salt in your everyday diet can quickly add up if your diet is composed primarily of packaged foods. People with high blood pressure need to be particularly cautious about their intake of salt, and may find that a diet of processed foods goes beyond the level of salt intake recommended by their doctors.

Many packaged foods also wind up being contaminated by their packaging. Most canned foods are packed in cans made with **lead** solder, which can leach into the food, doubling or tripling its original lead content. Continuous low-level exposure to lead has been found to produce permanent neuropsychological defects and behavior disorders in children, including low IQs, short attention spans, hyperactive behavior, and motor difficulties. Low-level exposure may also be carcinogenic, mutagenic, and teratogenic. There is no demonstrably safe level for lead. Pregnant women and children should be especially cautious about eating canned foods.

Other packaged foods (such as cheese spreads, whipped cream or no-stick pan coatings) come in **aerosol** or pressurized containers. When inhaled, aerosols are known to cause lung diseases in normal use and death in high concentrations. Because aerosol spray containers are a form of packaging, aerosol propellants are not considered an "additive" and their safety is not discussed in literature on food additives. Again, it's up to us to learn about these food contaminants and limit our intake.

Plastic is frequently used to package processed foods, and many times can leach into the food, with unknown health effects.

For most food products, federal law requires that nearly all food additives be listed on the label. Some packaged foods, however, may contain ingredients already containing additives, and these first-generation additives rarely make it to the label. Only the final ingredients used in a processed food must be listed, but not any of the ingredients found in the ingredients. Therefore, ham containing nitrites may be used, or shortening preserved with **BHA** and **BHT**, and only "ham" or "shortening" will appear on the product label.

WHAT YOU CAN DO

In response to our concern about the safety of food additives, more and more "natural" foods are introduced to the marketplace every day. Ironically, though, the growth of the industry has resulted in much confusion and some downright deceit regarding what is *really* natural. Let's try to clarify the problem, and find out the answer.

If a label says "100% Natural," there should be *no* artificial colors, flavors, preservatives, or other synthetic additives in the food products at all. You have to be very careful, however, because there is no legal definition of the word "natural" as it appears on food labels. Because of the large demand for natural products, many large corporations are now marketing their unnatural products in a way that misleads you into thinking they are natural. Don't be confused by pictures of farms on the label, or big letters announcing the absence of one particular additive. Read the ingredients carefully, and if it contains ingredients that you don't recognize as a food, don't buy it.

Don't be discouraged; just be careful. There *are* some additive-free foods at your supermarket. Check labels at your supermarket and at your natural food store—and compare prices. Some large corporations are actually removing artificial ingredients from selected items, and these are available at a lower cost than some natural brands with identical ingredients. On the other hand, supermarkets with natural food departments frequently sell the same foods found in natural food stores at greatly inflated prices.

While some truly natural food companies are especially careful about packaging, many products are still wrapped or contained in plastic, which may be absorbed into the food. Most canned natural foods, however, come in "lead-free" cans and more and more conglomerate food producers are also switching to lead-free cans, so look for them at the supermarket as well as at the natural food store. There are two basic types: lead-free soldered and one-piece aluminum. Once you know what to look for, they are very easy to identify.

Soldered cans have a top and bottom rim with a side seam that peeks out at the edge of the label. A lead-soldered can has an obvious, protruding side seam, and traces of solder may also be visible on the outside. If you run your finger around the top or bottom of the can near the rim, you'll be able to feel the lumpy seam of a lead-soldered can. These you should avoid. A *lead-free* soldered can has a flat, thin, neat seam with a characteristic narrow black or blue line running down the middle. You will recognize such a can as different from the lead-soldered variety right away. The lead-free aluminum can is also distinctive. Formed from one piece of metal, it has a rounded bottom, a rim at the top, and no seam.

In all fairness, I must mention that more and more conglomerate food producers are also switching to lead-free cans, so look for them at your supermarket, too.

FRUITS, VEGETABLES, AND OTHER PLANT FOODS ⊖

Are you one of the 77 percent of Americans (as reported by a Harris opinion poll) who believes that pesticides in food presents "a serious hazard"? If you are, your suspicion is increasingly confirmed by substantial evidence growing daily.

More than 200 million tons of **pesticides** containing more than 1000 active ingredients are used annually on croplands in California alone. In independent laboratory tests done in 1983 for the

Natural Resources Defense Council (NRDC), *44 percent* of the produce tested had detectable residues of 19 different pesticides. Almost half had residues of up to four different pesticides.

Three major classes of chemicals are used as pesticides on food: **organochlorines, organophosphates,** and **carbamates.**

Organochlorine insecticides work by attacking the central nervous system of the agricultural pest, resulting in convulsions and death. Many organochlorines cause cancer, birth defects, and genetic damage in animals. In addition, they are stored in our body fat, and can accumulate to high levels over time. Although these pesticides were used predominantly in the 1940s and 1950s, many are still in use today. Three organochlorines have been banned: DDT in 1972, aldrin in 1974, and endrin in 1979. However, these poisons can remain in our soil and water supplies an average of 50 to 75 years, adding dangerous contaminants to our food, without offering any positive agricultural benefit. Endrin, though banned for general use, is still allowed on apple and wheat crops. In the NRDC study mentioned above, DDT residues were found in more samples than any other pesticide, *eleven years* after the chemical was banned from use.

Organophospate insecticides are the most widely used insecticides today. Unlike organochlorines, organophosphates break down into harmless chemicals within several weeks, so they do not pose such a long-term threat. In pest protection, they work by blocking the action of the chemical that aids in the transmission of nerve impulses; as a result, those impulses are transmitted continuously, leading to convulsions and death. Some organophosphates have been found to cause genetic damage in animals, and others are believed to cause serious changes in brain activity.

Carbamates are also widely used today. Their toxicity and persistence lies somewhere between the organochlorines and organophosphates. One of the carbamates is suspected of causing birth defects, and a review of its safety has only just begun.

Two federal agencies hold responsibility for protecting us from pesticide residues in food: the Environmental Protection Agency (EPA) and the Food and Drug Administration (FDA). Under the Federal Insecticide, Fungicide, and Rodenticide Act (FIFRA), the EPA requires manufacturers of new pesticides to submit information about which crops it will be used on, their effectiveness and toxicity, and the nature and levels of residues. The EPA then sets a tolerance for each pesticide that establishes the amount allowed to remain on food crops.

This sounds like a pretty good system, but it has several flaws. Many pesticides now in use were granted tolerances *before* safety tests were required to determine whether the pesticide causes cancer, birth defects, genetic damage, or reproductive disorders. A study done by the National Academy of Sciences reports that 64 percent of pesticides now in use have not even been minimally tested for their toxic effects. Very little information is available on the potential long-term health effects, the possibility of synergistic effects resulting from exposure to more than one pesticide, and the range of individual sensitivities among the human population. Of the 19 pesticides detected in the NRDC survey to leave residues on produce, eight are suspected of causing cancer, five may induce birth defects, and the others could produce genetic mutations. But, for some of the pesticides used on food, no information is publicly available to enable us to assess for ourselves the possible health hazards.

The EPA sets acceptable pesticide residue tolerances for each individual crop or food type on which the pesticide will be used. The same pesticide might have a different tolerance for each separate fruit and vegetable. Tolerances are determined by considering both the safety of the pesticide and how much the EPA estimates you will eat of that food over the course of your lifetime, based on an estimate of how much an average person eats of that particular food in a year. But currently, new tolerances are set in accordance with a household survey taken in 1966, despite the fact that we have clearly increased our intake of fresh fruits and vegetables since that time. The EPA assumes each of us eats only about a *half a pound per year* of almonds, artichokes, avocados, barley, blackberries, blueberries, cantaloupes, eggplants, filberts, figs, garlic, leeks, mushrooms, nectarines, parsley, papayas, pecans, plums, radishes, rye flour, summer squash, and walnuts. Supposedly, we eat only about 1½ pounds of broccoli, 7½ pounds of carrots, 33 pounds of oranges (that's only about 66 oranges in 365 days), 20 pounds of lettuce, 84 pounds of potatoes, 3 pounds of strawberries, and 44 pounds of tomatoes. These figures make me wonder if the EPA has read the *Dietary Guidelines for Americans* recommended by the U.S. Department of Agriculture. I exceeded my year's quota for mushrooms at dinner last night.

Under the Federal Food, Drug, and Cosmetic Act, the FDA tests samples of food crops for pesticide residues. Again, this sounds good, but the tests commonly used by the FDA can only detect 107 of almost 300 pesticides that have established tolerances. As many as 20 pesticides suspected of causing cancer that are in the food supply are not even monitored by the FDA.

Further, although the monitoring system seems well intended, some highly toxic pesticides are found in our food supply that are actually banned for use in the United States. Organochorines such as **DDT, 2,4,5-T, aldrin, dieldrin, heptachlor, chlordane, endrin,** and **BHC** are illegal here but are shipped to other countries for use. They then return to us on imported foods—for example, coffee, chocolate, bananas, rice, tea, sugar, and tapioca. During the winter months, summer-season vegetables in supermarkets may be imported from Mexico, where pesticide use is not as tightly controlled as in the United States. These include zucchini, summer squash, yellow squash, garlic, string beans, okra, asparagus, bell peppers, cucumbers, eggplant, green peas, snow peas, and tomatoes. Even seasonally, if American crops fail or supplies aren't sufficient, green onions, radishes, parsley, corn, brussels sprouts, spinach, loose-leaf lettuce, watermelons, cantaloupes, honeydew melons, pineapples, and strawberries may also be imported.

Horrible as the pesticide problem is, it is not the only chemical concern we have regarding fresh produce. Bananas, pineapples, guavas, mangos, and other tropical fruits from Hawaii are fumigated with toxic substances before being allowed onto the mainland.

Also, another category of chemicals is introduced for "cosmetic purposes," to make our produce more attractive to the eye. Some oranges are dyed orange with the coal-tar dye Red No. 32, known to cause cancer in laboratory animals. And, before green oranges are dyed to look ripe, they are "degreened" with ethylene gas to remove the chlorophyll from the rind. The colorless oranges are then tinted by passing through a vat of hot dye. Almost all Florida citrus that reaches the marketplace before January 1 has been treated. Other produce that may be dyed are some "new" potatoes (regular potatoes dyed red) and some red yams (actually dyed sweet potatoes).

Concerns about packaging materials are by no means limited to processed foods. Citrus fruits are generally wrapped in papers treated with a potent fungicide. The first time I smelled an orange without this fungicide I was shocked to find that the smell I had associated with "citrus" all these years was not citrus at all; real citrus has a lovely, delicate fragrance, much different from the biting chemical odor of fungicide. With fungicides, you not only may be harmed by eating the food, but you could be affected by smelling it, too.

Some potatoes and onions are treated with **maleic hydrazide,** a potential human carcinogen, to inhibit sprouting and make old onions and potatoes appear fresh. It really doesn't inhibit sprouting at all, but causes the onions and potatoes to sprout *inside* instead of outside where you can see it! I stopped buying supermarket potatoes when I cut a half dozen open one day and they were all sprouted inside.

Carrots, oranges, lemons, limes, apples, pears, plums, peaches, melons, parsnips, tomatoes, green peppers, rutabagas, turnips, cucumbers, grapefruits, and tangerines may all be coated with paraffin, a petrochemical wax that may contain carcinogenic impurities.

Mushrooms might be grown without pesticides, but they are frequently fumigated with formaldehyde as a preservative.

Here's a final depressing warning: There is no way to tell what sort of chemical contaminants may be present in any supermarket produce.

WHAT YOU CAN DO

If you do have a local natural food store, please go there and support their effort to provide less contaminated *organic* foods. Even if they do sell some commercial produce, the produce people at a natural food store are much more likely to know where the produce came from and how it is produced than will supermarket personnel. Yes, organic food is more expensive, but you could well balance out the difference in health-care savings. Organic food is so important to me that I drive to the nearest natural food store (45 minutes away) once a week just to buy organic food.

We run into semantics problems in the produce department as well as in the packaged food department. Just because a sign says "organic," doesn't mean it's really organic. A few states actually have laws legally defining this term, but in most, the word is used pretty loosely. Look for produce approved by local certification programs or that the store can verify its production. Actually, you can sometimes recognize organic food by how it looks. Organic fruits and vegetables are often smaller than commercial produce, slightly misshapen, and uneven in color. But organic food tastes wonderful and eventually your eye begins to see the natural variety of these foods as beautiful, while paraffin-shiny apples or dyed oranges come to look totally unappetizing. Ask your local store what they know about their produce. If it's really organic, the produce people will be happy to talk with you and tell you why. And even if your natural food store carries no organic produce—it might be hard to come by in your area—they might have organic grains and legumes.

A variety of organic food is also available by mail order, at relatively reasonable prices. It's generally shipped right from the farm, so you can find out exactly how it's grown. Staying stocked up by ordering fresh produce may be a lot of trouble, but you might at least order organic grains, beans, and dried fruits in bulk.

If supermarket produce is your only option, the Center for the Biology of Natural Systems at Washington University in St. Louis, Missouri, recommends washing produce with liquid kitchen detergent to remove pesticide residues. While it will not remove all the pesticide residues, they report a removal rate of about 40 percent. I imagine liquid soap would also work, but whatever you try, make sure you rinse the produce thoroughly to remove the detergent or soap residue.

Removal of the outer portions of leafy vegetables can also reduce pesticide exposure. The outer leaves of both cabbage and lettuce can have 31 times the amount of residue as the inner leaves.

A true alternative to the hit and miss of buying produce is to grow your own produce at home. If you live in the country or in a suburb, you may already have a vegetable garden. Even in the city, you can grow your own in community gardens, on rooftops, in window boxes and balcony planters. Indoors, you can grow vegetables without soil in hydroponic hothouses. At the very least, you could try sprouts or mushrooms in a dark cupboard. But, if you garden, please learn organic gardening methods instead of using chemical pesticides and fertilizers. Your local library should have information on organic gardening methods, and there are many mail order companies catering to this popular field.

MEAT, POULTRY, MILK, AND EGGS ⊜

You're going to need a strong stomach to read this section—be warned! After you read this, you may become a vegetarian forever.

Most meat, poultry, and animal products on the market today come from factory farms that raise animals as "biomachines." The animals are confined in dark, crowded quarters, fed a diet high in drugs and chemicals, and low in nutrients, and they are deprived completely of exercise and fresh air. The results are highly stressed, deformed, and diseased animals, which hardly qualify as high-quality foods.

Many of the animals offered to us to eat in today's market are not even the same kinds of animals our grandparents ate. Thanks to the modern science of genetic manipulation, we are sold animals that grow faster (for efficient, high-yield production), but are of poorer quality. Hens are laying more eggs than ever before, but these eggs are smaller, with more white and less yolk, and they are paler and more watery than eggs from barnyard chickens.

The greatest source of contamination in factory animals comes from their feed. Farmers want to feed their animals something cheap that will make them grow heavy fast. Although some animals are fed soybeans, corn, barley, and other grains, others may be fed ground cardboard, old newspapers, sawdust, and recycled animal wastes. Researchers are even studying human sewage for ways to process it into animal feed.

Commercial feed may contain growth-stimulating hormones, coloring agents, fungicides and pesticides, drugs and medicines to treat diseases, and flavoring agents to make it more appealing to the animals. More than 40 percent of antibiotics produced in the United States are used as animal feed additives. Fully 100 percent of all poultry, 90 percent of pigs and veal calves, and 60 percent of all cattle are fed antibiotics as a regular part of their diets. And 75 percent of all hogs are given sulfa drugs. In all, more than 1,000 drugs and another 1,000 chemicals are approved by the FDA for use in feed. And many of these substances wind up in our bodies via burgers, chicken dinners, and bacon breakfasts.

Many chemicals in the feed alone are likely to leave residues in meat and poultry. Organochlorine pesticides used on feed grains can accumulate to very high levels in animal fat. The General Accounting Office of the U.S. government has identified 143 drugs and pesticides that are likely to leave residues in raw meat and poultry. Of these, 42 are known to cause or are suspected of causing cancer, 20 can cause birth defects, and 6 can cause mutations.

Calves raised for "milk-fed" veal are fed a "milk replacer" formula containing dried skim milk, dried whey, starch, fats, sugar, mold inhibitors, vitamins, and antibiotics. No iron is allowed in their diet, because iron discolors their meat. Farmers must choose exactly the right time to take them to the slaughterhouse—when they've reached a minimum weight, but before they die from anemia. Typical farms lose from 15 to 50 percent of their calves to disease.

Many chickens, ducks, and turkeys suffer from retarded growth, eye damage, blindness, lethargy, kidney damage, disturbed sexual development, bone and muscle weakness, brain damage, paralysis, internal bleeding, anemia, and physical deformities as a result of vitamin deficiencies. Chicken feed contains artificial color to give a healthier color to undernourished skin. After slaughter, carcasses are washed in chlorine solution to kill salmonella bacteria. Chlorine residues may remain as the birds go to market.

Egg-laying hens also get artificial color in their feed—to color the pale yolks of their eggs. Some egg producers add chemicals to the feed, which passes through the digestive tract and into the manure, where it kills fly larvae. Of course, these chemicals probably also end up in the eggs. Arsenic is frequently added to feed to speed maturation and stimulate egg production. These speedy hens may produce more eggs in a shorter amount of time, but a barnyard hen lays eggs for 15 to 20 years, and factory hens last only about a year and a half before they are made into soup and other processed foods.

U.S. Department of Agriculture (USDA) meat and poultry inspection programs were established in 1906, before modern farming methods increased production to more than 10 billion pounds of animal carcasses each year. An inspector is allowed about 3 seconds to examine each carcass. Random samples of residues from carcasses are taken and sent to a laboratory for analysis. By the time the results are back, the meat is already in the stores. The USDA does not even test for 97 drugs and pesticides likely to leave residues or 24 known or suspected carcinogens. Fully 17 of the chemicals in meat and poultry are suspected of causing birth defects.

Despite inadequate facilities, however, USDA inspectors do ferret out hazards. Animals are examined before slaughter for obvious signs of disease or abnormality. Each year about 116,000 mammals and nearly 15 million birds are condemned before slaughter. After killing, another 325,000 carcasses are discarded and more than 5.5 million major parts are cut away because they are determined to be diseased. Shockingly, 140,000 tons of poultry is condemned annually, mainly due to cancer. The diseased animals that cannot be sold are processed into . . . animal feed.

WHAT YOU CAN DO

Considering current growing practices, it's not surprising that diets heavy in meat and animal fats have been associated with higher incidences of heart disease, colon cancer, stroke, and other degenerative diseases.

Many Americans eat much more protein than they actually need. Without ever eating animal products we can fulfill all our protein needs with amino acids from grains, legumes, nuts, and vegetables, if we choose. Even the "Dietary Guidelines for Americans" recommends reducing consumption of meat.

If you continue to keep meat in your diet, check your natural food store for meat and poultry from sources whose animals have been pasture grazed and raised without hormones or other chemicals to stimulate or regulate growth or tenderness, and without drugs or antibiotics. Also look for game that has been hunted in its natural habitat and contaminated only by those chemicals found in the natural environment. In some states, these flesh foods may be referred to as "organic," but, again, in most areas this term has no legal definition. Your local natural food store might have fresh or frozen chicken, beef, or lamb from farms raising their stock organically, and a whole variety of uncontaminated wild meats are available by mail, including buffalo, elk, bear, moose, reindeer, hippopotamus, llama, mountain lion, beaver, and other exotic animals.

If your only source of meat is the supermarket, you can reduce your exposure to toxic substances buy avoiding liver, sweetbreads, and other organ meats, which contain the greatest concentrations of toxics. Remove fat before cooking other cuts, as toxics tend to accumulate in the fat, rather than in the muscle flesh. And ask your butcher for a fresh cut wrapped in butcher paper, rather than meats and poultry in a heat-sealed plastic package.

Buy eggs from your natural food store. They will carry eggs laid by free-running hens that are not fed antibiotics or stimulants. These growing practices will be stated on the carton. Check your supermarket, too, as many supermarket chains are realizing that uncontaminated eggs sell very well. "Fertile" eggs are often produced with natural methods, but the word "fertile" alone on the carton does not guarantee the eggs are natural; it simply means that the eggs can be hatched into baby chicks. Some people believe that fertile eggs have added health benefits.

Milk can contain residues of any of the chemicals or drugs used in the feed. All milk sold in the United States today contains pesticide residues; no milk is free from them because all grazing lands are still contaminated with residues of banned pesticides such as DDT. Fortunately, the contaminants tend to accumulate in the fat molecules, so the lower the fat content of the milk product, the lower the contamination. Nonfat milk would have very low concentrations; butter would be highly contaminated.

Most milk on the market has been pasteurized, homogenized, and fortified. Pasteurization is simply a heating process that destroys bacteria. Homogenization mixes milk under pressure to reduce fat particles to a uniform size in order to improve taste, color, and the tendency to foam. Neither of these processes adds chemical contaminants, but it is very rare to find pasteurized and homogenized milk that has not been fortified with vitamins A and D, a process that adds propylene glycol, alcohols, and BHT.

If possible, buy milk in glass bottles. Some local dairies will deliver milk in glass right to your door. Milk in paper cartons would be a second choice. Avoid milk in plastic containers because it can absorb plastic, which then goes straight into your body when you drink.

Some people interested in health foods like to drink "raw" milk. It has its advantages, mainly that it has not been processed in any way and therefore contains no additives. However, the safety of this milk is a controversial subject, because raw milk has been linked with a number of cases of salmonella poisoning (explained later in this chapter). Health food advocates say that raw milk is more healthful because the pasteurization process destroys vitamins and enzymes. According to an article in the *Journal of the American Medical Association,* pasteurization affects only three vitamins for which milk is a minor source: thiamine, B12, and C, and none of these losses exceeds 10 percent.

I prefer raw milk because it is closer to the milk that actually comes out of the cow. If you are more concerned about salmonella than about the lost nutrients, you could boil the raw milk yourself and you would have a much purer product than standard pasteurized/homogenized/fortified milk.

You could also try soy milk, which you can make yourself from organic soybeans, or buy at your natural food store.

FISH AND SEAFOOD ⇔

Fish and seafood commonly contain high levels of the water pollution contaminants present in the area from which they are taken. Fish can concentrate pollutants to levels up to 2000 times more than the surrounding waters. The two water pollutants of greatest concern found in fish and seafood are toxic **mercury** and carcinogenic **PCBs**. Radioactive materials are also found in fish and seafood harvested near marine dump sites.

Once the seafood is harvested, chemical preservatives may be used to keep them fresh. **Sulfites** are sometimes used on clams, lobster, crab, scallops, and shrimp. **Sodium benzoate, polytrisorbate**, and **polyphosphates** may also be used.

WHAT YOU CAN DO

Select species of fish that spend most of their lives in deep waters, far out at sea and away from human pollution: herring, sardines, anchovies, small salmon (pink, coho, sockeye, and Atlantic), scrod, hake, haddock, pollock, mackerel, pompano, red and yellowtail snapper, striped bass, butterfish, squid, octopus, and tilefish. Avoid coastal and freshwater fish, which are more likely to have encountered polluted waters.

Shellfish are a little more difficult to buy. Since all the commonly eaten shellfish live in coastal waters and their natural function causes them to pass a great deal of water through their systems, they are natural collectors of high levels of contaminants. It's best to eat these as infrequently as possible.

If you like to eat fish as a regular part of your diet, find a local fresh fish market where you can become aquainted with the manager and find out the source of the fish you buy and if it has been treated with chemicals.

SWEETENERS ⬭

From the Manufacturers' Warning Labels:

> Use of this product may be hazardous to your health. Contains saccharin, which has been determined to cause cancer in laboratory animals.

Let's face it. We all know sweet things "aren't good for us," but we're going to eat them anyway. Right? So all we have to do is choose the sweet that is least harmful, and eat it in moderation.

To begin with, let's discuss the most dangerous sweets: artificial sweeteners.

For many years saccharin has been widely used as an artificial sweetener, despite the fact that it can cause cancer in humans. After several attempts to ban the substance, the FDA has now scheduled it for removal from the market in 1987.

Aspartame (NutraSweet, Equal) is the new favorite; it's a "natural" sweetener made of phenylalanine and aspartic acid, containing "nothing artificial." In the body, these naturally occurring substances break down into the same amino acids found in any protein food. Sounds harmless, but it isn't.

The problem with aspartame lies in overconsumption and the fact that phenylalanine alone (without its companion amino acids) is not a normal part of the diet. Large doses of phenylalanine are toxic to the brain and can cause mental retardation and seizures in people with phenylketonuria (PKU), a genetic disorder. For others, the sweetener may cause chemical changes in the brain that could contribute to headaches, depression, mood swings, high blood pressure, insomnia, and behavior problems. In addition, it could cause your appetite control center to malfunction, so your diet drinks could be causing more harm than good. Aspartame may also cause birth defects, such as mild retardation, and is not recommended for use by pregnant women.

Because aspartame is found in so many products, it is very easy to overdose without realizing it. A child meets the FDA maximum safety limit by drinking only 5 cans of diet soda per day; a 150-pound adult would exceed the limit by drinking 16 cans. This sounds ridiculous (how many people drink 16 cans of diet soda each day?), but when you take a vitamin pill with aspartame, eat your breakfast cereal and hot cocoa with aspartame, have some aspartame-sweetened gelatin and a soft drink for lunch, chocolate pudding with aspartame for dinner dessert, and maybe another soda, it adds up very quickly. Children could easily consume twice the FDA limit every day, and possibly suffer learning impairment and behavior problems. Part of the problem with the current labeling for aspartame is that the actual amounts used do not have to be listed, so you really have no idea how much aspartame you are consuming.

WHAT YOU CAN DO

So when the sweet tooth strikes, what do you do?

Teaspoon for teaspoon, eating white sugar (sucrose) in moderation as part of a diet that is based on fresh whole foods is probably safer than using any artificial sweetener. Moderation is the key word here, for a high-sucrose diet can cause significant nutritional deficiencies, lowered resistance to disease, tooth decay, diabetes, hypoglycemia, heart disease, ulcers, high blood pressure; it can stimulate your appetite and make you fat. Steer away from processed foods with high amounts of hidden sucrose (it has many names, including sugar, corn sugar, corn syrup, dextrose, and glucose syrup). Buy unsweetened products, such as breakfast cereal, and add sugar with discretion. Here's where reading labels can really pay off.

As a food, refined white sugar is highly contaminated, having been sprayed with multiple pesticides, processed over a natural gas flame, and chemically bleached. With this in mind, and considering the health effects of sugar, you may want to choose one of the sweeteners offered at your natural food store that are less contaminated and owe their sweetness to sugars other than sucrose.

When one thinks of a "healthy" sweetener, the first one that comes to mind, of course, is honey. It is composed of glucose and fructose and is generally the least chemically contaminated sweetener because bees don't come back to the hive if they've been exposed to pesticides. Choose a light-colored honey such as clover, star thistle, mountain wildflower, or orange blossom for a general-purpose sweetener. These are available at every supermarket. Darker honeys have some very interesting flavors, but generally tend to overpower other flavors when used in recipes as a neutral sweetener. You can replace honey in any recipe calling for sugar by using ½ to ¾ cup honey for each cup of sugar, and by reducing the liquid in the recipe by ¼ cup for each ¾ cup honey used.

Fructose (fruit sugar) is a popular sweetener, now used in many "natural" products as a sugar replacement. It is the sweetest sugar, so you get more sweetness per calorie and you can use less. The problem with fructose is that as a commercial product it is about as close to the natural sugar found in fruits as white sugar is to the sugar cane. Like aspartame, it is an isolated form of a chemical component usually found in combination with other components that help our bodies assimilate it. Another problem with fructose is that it is often made from corn syrup and can contain up to 55 percent sucrose, so if you're trying to avoid the latter, fructose won't help much. If your natural food store carries *pure* fructose, you might be interested in trying it.

Maple syrup is nothing more than boiled-down maple tree sap —real maple syrup, that is, not the artificially flavored corn syrup called "pancake syrup." It's not hard to find. If your supermarket doesn't carry it, your natural food store will. Check labels carefully, however, and buy a Canadian brand or one marked "no formaldehyde." While maple trees are grown without fertilizers or pesticide sprays, American law permits formaldehyde pellets to be inserted in the trees to increase the flow of sap, and it doesn't require that this practice be noted on the label. Maple syrup is a sucrose sweetener, and has all the same health risks as white sugar, so you may want to limit your intake to a trickle over your Sunday waffle, rather than as an all-purpose sweetener.

Two wonderful, delicious, and easy-to-use sweeteners are date sugar and fig syrup. Because they are made from real foods, they are among the few sweeteners that have some nutritive value. Date sugar is made by dehydrating dates and grinding them up into a rather coarse, granulated-type sugar. Fig syrup is made simply by boiling figs in water. Although date sugar will not dissolve very well in your cup of coffee, it works very well, substituted cup-for-cup, in any baked goods recipe that calls for brown sugar. Substitute fig syrup, cup-for-cup, in recipes that call for molasses. Here's a recipe in which you can try them both:

Gingerbread Cookies
¼ *cup butter*
½ *cup date sugar*
½ *cup fig syrup*

3½   *cups whole wheat flour*
1   *teaspoon baking soda*
¼   *teaspoon ground cloves*
½   *teaspoon ground cinnamon*
2   *teaspoons powdered ginger*
¼–⅓ *cup water*

1. *Blend together butter, date sugar, and fig syrup.*

2. *Sift together flour, baking soda, and spices.*

3. *Combine butter and flour mixtures together alternately with water. You probably will have to use your hands to mix near the end. For softer cookies, add more water.*

4. *Roll out dough to about ¼ inch and cut out shapes. Cookies can be decorated with bits of dried fruits.*

5. *Bake at 350° for 8 to 10 minutes, or until lightly browned.*

You will find other natural sweeteners at your natural food store, but they are more difficult to use and almost impossible to substitute in regular recipes. Two of these sweeteners are made from barley malt. The syrup is less than half as sweet as sugar, and the powder is *2000 percent* sweeter than sugar (when you are measuring 2 teaspoons of barley malt powder to replace 1 cup of sugar, it is easy to get too much). Rice syrup is best used in place of jam on toast or as a syrup on pancakes, as it is only 20 percent as sweet as sugar.

Just because these sweeteners are more natural doesn't mean they are foods. They can be just as fattening as white sugar, so use them sparingly—for your health and your waistline.

FOOD POISONING ☜

Not all harmful contaminants in food are chemical or added by human beings. Nature has its own toxins that occasionally show up, too.

Salmonella, Clostridium perfringens, Staphylococcus, and *Clostridium botulinum* are little bacterial organisms that multiply by dividing. They can contaminate all types of foods, causing stomach upset, abdominal pains, diarrhea, headaches, and, in extreme cases, death.

Salmonella is one of the most common causes of food poisoning. Fortunately it is not often fatal—more than two million cases of *salmonella* poisoning are believed to occur in the United States each year. Salmonella is most commonly found in raw meats, poultry, eggs, milk, fish, and products made from these foods. Chocolate, yeast, and spices have also been known to carry this bacteria. Symptoms of salmonella poisoning are fever, headache, diarrhea, abdominal discomfort, and vomiting. These symptoms appear in 24 hours, and most people recover in 2 to 4 days. Children under 4, elderly people, and people already weakened by disease could become seriously ill from these bacteria.

Clostridium perfringens is more widely distributed over the earth than any other disease-causing microorganism—in the soil, dust, on food, and in the intestinal tracts of humans and other warm-blooded animals. Large numbers of these bacteria will cause diarrhea and abdominal pain about 12 hours after the contaminated food is ingested.

Staphylococcus organisms are in your respiratory passages and on your skin. They usually enter food by way of human contact and can multiply in meats, poultry and egg products, egg, tuna, chicken, potato or macaroni salads, cream-filled pastries, and sandwich fillings. If staph germs are allowed to multiply to high levels, they form a toxin that cannot be boiled or baked away. Symptoms of poisoning include diarrhea, vomiting, and abdominal cramps that occur 1 to 7 hours after eating, and subside within 24 to 48 hours.

Clostridium botulinum poisoning is rare—fortunately, for it is often fatal. Spores are found throughout the environment and are harmless until they divide in the proper environment and produce a poisonous toxin. Symptoms begin 12 to 36 hours after eating and include double vision, inability to swallow, speech difficulty, and progressive paralysis of the respiratory system. If botulism is suspected, call a doctor immediately—medical help must be obtained at once.

Another kind of poisoning comes from aflatoxin, the product of two common molds, *Aspergillus flavus* or *Aspergillus parasiticus*.

Aflatoxin has been detected in corn, wheat, barley, rice, figs, grain sorghum, cottonseed, certain tree nuts (such as Brazil nuts) and peanuts, and is known to cause hepatitis and liver cancer in humans.

WHAT YOU CAN DO

These bacteria can enter food in any stage of growing and processing. In some cases, food you buy may already be contaminated, but you are the one who is responsible for safe handling practices from the supermarket or natural food store to the dinner table.

When buying foods, don't buy any foods in packages or cans that are outdated, broken, bent, or leaky. Especially avoid bulging cans, as these could contain *Clostridium botulinum.* Make sure meat, poultry, and frozen foods are kept cold. Buy them last, so they don't warm up or defrost in your cart while shopping. Make the grocery store your last stop while shopping so you can go right home with perishables. Put meat and poultry in the refrigerator at 40° F or the freezer at 0° F immediately when you get home. Also put frozen foods in the freezer right away. Refrigerating or freezing foods prevents any existing bacteria from multiplying.

Thaw frozen foods in the refrigerator before cooking, or cook them while still frozen. To thaw foods quickly, use a watertight plastic bag submerged in cold water. Never thaw frozen food uncovered on a kitchen counter at room temperature. There are bacteria everywhere that could quickly multiply.

When preparing foods, take care that cross-contamination does not occur. After cutting up raw meat, poultry, or fish, wash your hands, knife, and cutting board thoroughly with hot water and soap before preparing other foods with the same equipment. Bacteria from the meat, poultry, or fish can easily be transferred to other foods and then multiply later, when conditions are favorable.

If you have cuts or sores on your hands, use rubber gloves when preparing foods to keep both the bacteria from the food out of your sores, and the bacteria from your sores out of the food.

Because bacteria can grow rapidly at room temperature, don't keep cooked main dishes at room temperature for more than 2 hours before serving, and refrigerate leftovers as soon as the meal is over.

Pets are also carriers of bacteria. After handling pets, wash your hands before preparing food. Keep pet-feeding dishes, toys, and bedding out of the kitchen. Don't let pets touch food, utensils, or work surfaces where food is prepared.

Most bacteria are destroyed by heat, so always use a meat thermometer and cook foods thoroughly. Also, thoroughly heat leftovers before serving, don't just "warm them up."

The most common cause of botulism poisoning is home canning. Be sure to follow proper procedures to sterilize and seal jars, and ensure that the food is thoroughly cooked. Botulism can also develop in foods that are in an atmosphere free from oxygen and slightly warm. This sounds almost impossible, but such situations may happen frequently—for instance, when you leave a leftover baked potato wrapped in foil overnight at room temperature or leftover vegetables sauteed in butter or oil.

There is little we can do to protect ourselves from aflatoxin, since it can be present and invisible in foods when they are purchased. If nuts are a part of your diet, store them in a cool, dry place, and take care not to eat discolored or moldy nuts. Otherwise, all you can really do is avoid those foods most likely to contain aflatoxin.

FOOD IRRADIATION ☞

As I write this, irradiated food is not yet on the supermarket shelves, but by the time you read this, it may be.

The irradiation process itself is simple: Fresh or frozen foods are moved on a conveyor system past a radiation source that zaps it with gamma rays. The food industry is in favor of irradiation because it will increase the shelf life of perishable foods and might also prevent food-borne diseases.

The safety of these foods is very controversial. Although radiation similar to X rays is used to preserve foods, the foods themselves do not become radioactive. Some say they are absolutely safe; others suspect a regular diet of irradiated food may cause leukemia, other forms of cancer, and kidney disease. At this point, we really don't know the possible health effects.

Irradiation can also alter the nutritional quality of food, destroying proteins and vitamins, although probably no more than would be destroyed in cooking. Still, the cooking of irradiated food would have far fewer nutrients than if a fresh food were cooked.

The present law requires that irradiated foods be labeled as such. The FDA is now reconsidering this ruling, and studying the possibility of reclassifying irradiation as a processing step, rather than a food additive, which would then eliminate the need for labeling.

I am not in favor of irradiation. Even if potential health effects are not yet proved, it can't be good for us to eat foods that have been so unnaturally altered. I just hope that irradiated foods will be labeled so that we know what is available to us and can have a choice.

COOKWARE ☕

Copper and **aluminum** cookware should not be used if the cooking surface that comes in contact with the food is made from either of these metals. Foods cooked in aluminum can react with the metal to form aluminum salts. Little research has been done on the amount of exposure we may receive from food cooked in aluminum cookware or its health effects, but aluminum salts from other sources of exposure have been connected with brain disorders such as dementia, Alzheimer's disease, behavior abnormalities, poor memory, and impaired visual-motor coordination. One British study showed that foods cooked in aluminum cookware may cause indigestion, heartburn, intestinal gas, constipation, and headaches.

Stainless steel cookware may also be harmful. It seems that if stainless steel cookware is scoured only once with an abrasive powder or steel wool, small amounts of highly toxic metals such as **chromium** and **nickel** may be dissolved into every food cooked in it thereafter. But if stainless steel cookware is always soaked clean and no abrasives are used, it should be harmless.

Pots and pans with "no-stick" finishes such as Teflon or Silverstone should not be used, period. They can scratch easily and contaminate foods with bits of plastic during cooking.

WHAT YOU CAN DO

So what's left? Glass, cast iron, porcelain-enamel coated cast iron or stainless steel, or terra cotta clay. Cookware made from all these materials is available at most department stores and kitchen shops.

For a no-stick finish, season cast iron cookware before using by covering the bottom of the pan with cooking oil and placing the pan on a warm burner or in a warm oven for one hour. Wipe out excess oil, leaving a thin film of oil in the pan. Each time you use it, the pan will become more seasoned, if you don't allow the food to stick. Maintain the seasoning by wiping the pan with a clean towel after cooking or rubbing the bottom of the pan with salt.

For an instant no-stick finish, heat the pan first, then add the oil, then add the food to the hot oil. If you're on a diet, spread a film of liquid lecithin over the cooking surface before using it. Lecithin is the active ingredient in no-stick aerosol cooking sprays, and can be purchased at your natural food store.

Line baking pans with parchment paper (available at most cookware stores) for no-stick baking and easy clean-up.

MICROWAVE OVENS ✍

The safety or danger of exposure to microwaves from microwave ovens has not yet been established one way or the other. There is some concern because even microwave ovens that are functioning perfectly emit microwaves. FDA safety standard limits allow microwave emissions of up to 1 milliwatt per square centimeter ($1mW/cm^2$) when the oven is purchased, and up to $5mW/cm^2$ after the oven has been in use.

Animal studies done by Bell Telephone Laboratories have convinced this company to establish tighter safety regulations than those required by the U.S. government. They recommend only "incidental exposure" from 1 to 10 milliwatts, and that daily exposure should not exceed 1 milliwatt. Russian safety regulations prohibit exposure at 1 milliwatt for more than one minute, and then only with

goggles to protect the eyes. Clearly, these regulations apply to microwave exposure well within the range that might be encountered while
using a microwave oven.

Workers who are subjected to exposure to microwaves on the
job complain of headaches, fatigue, irritability, sleep disturbances,
weakness, increased incidence of viral infections, low heart rates,
changed EEGs, and increased thyroid functions. There is also some
concern that microwaves may cause birth defects and a possibility that
they may also cause cancer.

WHAT YOU CAN DO

How about going back to cooking the old way, in the oven and
on top of the stove?

If you have a microwave oven, and you really love the conve-
nience of it, you may not have to give it up. *Consumer Reports* magazine suggests that you can minimize your risk by keeping a reasonable
distance from the oven while it is in operation (the farther, the better),
and try to operate and maintain the oven in such a way that will
minimize leakage. Make sure the oven door closes properly and that
no damage occurs to the hinges, latches, sealing surfaces, or the door
itself. Make sure that no soil or food residue accumulates around the
door seal, and avoid placing any objects between the sealing surfaces.
You might also want to have the oven tested every few years for
leakage.

BARBECUED AND BROILED FOODS ☞

Most barbecued and broiled foods contain benzo(a)pyrene, a
proven cancer-causing substance.

The cooking temperature, type of fuel used, and the fat content
of the meat all affect the amount of benzo(a)pyrene that is formed.
Cooking methods that create the most benzo(a)pyrene are charcoal
grilling and gas grilling, when the gas flame is *below* the meat. Grilling meats closer to the coals produces more of the carcinogen than
meats grilled farther from the heat source. Benzo(a)pyrene seems to

form most rapidly when a fatty meat is cooked over gas or charcoal close to the heat source for a long period of time.

WHAT YOU CAN DO

For some unexplained reason, broiling with a gas flame or electric coil *above* the meat produces *no* benzo(a)pyrene. So slather on the barbecue sauce and stick your meats under the broiler!

ALCOHOLIC BEVERAGES ⇔

The health effects of alcoholic beverages are well known. Alcoholism can cause heart disease, hepatitis, cirrhosis of the liver, decreased resistance to disease, shortened life span, nutritional deficiencies, cancer, fetal alcohol syndrome, brain damage, stroke, phlebitis, varicose veins, and a reduced testosterone level in males that causes sexual impotence, loss of libido, breast enlargement, and loss of facial hair. A bill presented to the House of Representatives suggests that labels on alcoholic beverages contain a warning: "CAUTION: The Surgeon General has determined that consumption of alcoholic beverages during pregnancy can cause serious birth defects. Alcohol can also impair driving ability, create dependency or addiction, and can contribute to other major health hazards."

A nonalcoholic danger found in most wines and beers is **sulfites**. Unfortunately for those who are sulfite-sensitive, the Food, Drug, and Cosmetic Act is not enforced when it comes to ingredient labeling for alcoholic beverages. Sulfites in wines have caused many unsuspected allergic reactions.

WHAT YOU CAN DO

There are good brands of additive-free wines and beers on the market; a few wines are even made with organic grapes.

For those who wish to avoid alcoholic beverages entirely, a number of delicious nonalcoholic drinks are available at your natural food store. Try one of the no-alcohol beers, a varietal grape juice made from wine grapes, or a sparkling apple cider or other fruit juice instead of champagne. Read labels carefully, though, to watch for sulfites. Because juice is a *food*, and not an alcoholic beverage, any sulfite in the product must be listed on the label.

Beware of "dealcoholized" wines. Because they begin as wines and then have their alcohol removed, they still may contain the same sulfites and other additives used in winemaking. The National Council on Alcoholism also warns that "nonalcoholic" or "dealcoholized" on a wine bottle or beer can does not necessarily mean completely free of alcohol. Under federal law, these beverages may contain up to 0.5 percent alcohol, without stating so on the label. This regulation appears to apply only to those fermented alcoholic products from which the alcohol has been "removed," and not to those, such as varietal grape juices, that have never been fermented.

VITAMIN AND MINERAL DIETARY SUPPLEMENTS

The only vitamins and minerals you need to watch out for are the heavily advertised, brightly colored, high-potency multivitamins with clever names. They may contain **BHA, BHT, artificial colors, artificial flavors, mineral oil, sugar, sulfites,** and **talc.** The brightly colored coatings may be made, not from food, but from **plastic;** those in gelatin capsules may be preserved with **formaldehyde.**

WHAT YOU CAN DO

First, if you are going to take vitamins, at least take a brand that is additive-free. Your natural food store is probably full of them, so you will have no problem finding them.

The more complex question is: Should you take vitamins and minerals taken from natural sources, or synthetic vitamins made from petrochemicals?

Or should you take supplements at all? Do you need them? This is a very controversial subject and the answer ranges from "No, you get adequate nutrients in your daily diet" to "Yes, you need mega-amounts to compensate for the nutrients missing in our modern food supply, our high level of daily stress, and our exposure to environmental chemicals." *Your* individual need probably falls somewhere between the two.

Whether to take natural or synthetic vitamins is also a controversial subject. Despite the fact that the two types have identical chemical structures, they have subtle biological differences and different levels of biological activity, which affect how much of the vitamin the body can actually use. Many reports indicate that natural vitamins, with their higher biological activity, are better utilized by the body than synthetic forms.

The only truly natural vitamins and minerals are those that come from foods. They are available as highly nutritious, concentrated foods (such as bee pollen), powdered concentrates of foods with the moisture and fiber removed (as in barley-leaf juice powder), or the isolated component of a food (wheat germ oil, for example). Natural food supplements you will find at your natural food store include alfalfa, barley-leaf juice powder, bee pollen, bone meal, brewer's yeast, chlorophyll, cod liver oil, desiccated liver, kelp, spirulina, and wheat germ oil.

Most so-called "natural" vitamins on the market today are either fortified (low-potency natural vitamins mixed with high-potency synthetic vitamins) or synthetic vitamins in a natural base (the label will say something like "in a natural base containing . . ."). For most people, either of these types of vitamins are perfectly acceptable.

CHAPTER 8

Clothing, Bedding, and Laundry Products ☞ ◉ ✐ ℓ

The nontoxic closet is filled with cool cotton shirts, warm wool jackets, sensuous silk dresses, crisp linen blouses, luxurious cashmere sweaters, silk stockings, cotton socks, and cotton or silk undergarments. For bed linens, you could choose from smooth cotton percale, warm cotton flannel, sexy silk, or elegant imported linen. Moths are discouraged from entering if the delicate scent of lavender or cedar is nearby.

Sound expensive? It needn't be. Natural fibers are becoming much more popular, and, although a $500 cashmere blanket is costly, it's possible to buy natural fibers at very reasonable prices. You probably already have some natural-fiber clothing in your closet. Go through your wardrobe and separate the natural-fiber clothing from synthetic clothing, wear natural fibers most of the time, and, when you shop, look for items made from natural fibers.

A greater danger from textile products comes not so much from the fibers themselves, but from the finishes that are applied and the chemicals we use to care for them. But there are safe alternatives to these too.

SYNTHETIC FIBERS ☞

The clothing and bedding found in a ''chemical closet'' is made from synthetic fibers—**nylon, polyester,** and **acrylic** are the most popular. Even though these fibers may look like the natural fibers they attempt to imitate, they are actually very soft thermoplastics (see Chapter 10) made from petrochemicals.

Very little research has been done on the possible toxic effects of wearing plastic fibers. Still, these are plastics, and they continuously give off minute plastic vapors as the fiber is warmed against your skin. Many people experience irritating skin reactions. While this certainly isn't life-threatening, anyone who has ever spent an entire day itching all over knows how annoying it can be.

Perhaps a more practical reason for not wearing synthetic fibers is that they just aren't very comfortable. None of them absorbs moisture very well, and they can leave you feeling hot, sticky, and clammy during warm weather. They're not much better for winter wear because they don't hold the heat of your body very well.

Plastic fibers also have maintenance problems. They tend to absorb oil from your skin and hold oily stains that can only be removed with specially developed synthetic detergents (I'll discuss those later). And static cling is a condition unique to synthetic fibers, caused by an electric charge created by the friction of the synthetic fiber against your body. Even more synthetic chemicals are used in fabric softeners and aerosol antistatic agents in an attempt to solve this problem.

WHAT YOU CAN DO

Natural fibers are the choice of many people simply because they are more comfortable and look fashionable. Cotton comes in many weights and textures, from crinoline to canvas. Linen fibers from the flax plant are especially natural, as they are usually left in their natural shade or bleached, rather than dyed. Silkworms spin silk that can be slinky or nubby. Wool has a variety of weights and natural shades, and can come from lambs, llamas, alpacas, rabbits, yaks, Kashmir goats, and camels. They are cool in the summer, warm in the winter, and very few chemicals are used in their manufacture.

The Textile Fiber Products Identification Act, passed in 1960, requires that each textile product be labeled with the generic names of the fibers from which it is made, so you will have no trouble finding items made from natural fibers.

FLAME-RESISTANT FABRICS ✍ 👆

The Flammable Fabrics Act was passed in 1953 to protect consumers from injuries and deaths associated with highly flammable fabrics. According to the Consumer Product Safety Commission, every year thousands are hurt in accidents involving fire by playing with matches, smoking while cleaning paintbrushes with combustible solvents, reaching over lighted gas burners, and falling asleep while smoking in bed. The act has been amended over the years and, at this point, standards have been set for general wearing apparel, carpets and rugs, mattresses and mattress pads, and children's sleepwear.

Each category has its own standards for allowed flammability. The standard for general-wearing apparel is easy to pass; it does not require testing of individual garments and bans only extremely flammable items. Children's sleepwear is another matter. Each individual garment design must pass strict performance standards, as even the type of fiber used in the thread can affect the garment's ability to burn.

There are three types of flame-resistant fabrics: (1) synthetic fabrics made from fibers for which the flame resistance is inherent in the fiber, (2) synthetic fabrics made from fibers to which flame-retardant chemicals are added during manufacture, and (3) fabrics to which flame-retardant chemicals are added after the fabric is woven.

Nowadays, most children's sleepwear is made from polyester, which is inherently flame resistant. Use of chemical flame-retardant finishes (such as TRIS, which was banned for use in children's sleepwear because of its carcinogenicity) have, in the words of an employee of the Consumer Product Safety Commission, "pretty much gone by the wayside." It's very unlikely that you would encounter a chemical flame-retardant finish on general-wearing apparel either.

If you are concerned about a chance encounter with a rare

flame-retardant chemical, it may be difficult to recognize. Manufacturers are not required to label garments as flame resistant, but they may identify their garments or fabrics as flame resistant as a selling point. But manufacturers *are* required to label flame-resistant fabrics with precautionary instructions to protect the fabric from treatments that may cause their flame resistance to deteriorate. A good clue for finding flame-resistant fabrics would be to look for labels warning against washing the fabric with nonphosphate detergents in hard water, washing the fabric with *soap,* or using bleach during laundering.

WHAT YOU CAN DO

Just out of curiosity, I looked up "polyester" in a textile textbook I have called *Understanding Textiles* by Phyllis G. Tortora to see what its combustibility is in relation to natural fibers. The book states: "Polyester shrinks from flame and will melt, leaving a hard, black residue. The fabric will burn with a strong, pungent odor. Some polyesters are self-extinguishing. *Melted polyester fiber can produce severe burns* [italics mine]." I am assuming that the polyester fibers used to make children's sleepwear are the self-extinguishing sort.

So what can you do if you are concerned about fabric flammability and want to ensure that your clothes are flame-retardant? First, you might like to know that although cotton and linen are relatively flammable, *silk and wool are self-extinguishing,* which means they will burn when a flame is present, but will go out when the flame is removed. The weave of a fabric contributes to its flammability—loose weaves are more flammable than tight weaves. The ideal, naturally flame-resistant fabric for children's sleepwear would be raw silk noil, a medium-weight, nubby fabric quite unlike the silk used for blouses. I have a scarf made from this fabric, and I just throw it in the washing machine with my cottons.

FABRIC DYES ✍

It's ironic that the most dangerous dyes are those used on natural fibers.

Direct dyes used on cotton fabrics contain highly carcinogenic **benzidines.** One of them, **dichlorobenzidine,** is very easily absorbed through human skin. Now you may wonder how the dye is released from the fabric onto the skin. Dyes are sometimes remarkably unstable. Look under the arms of an old shirt you have been wearing for a long time and see how the color has faded. Have you ever had the experience of wearing a dark-colored shirt under a light-colored sweater and the dark dye soaks right through the sweater when you perspire? If the dye is released, it can be absorbed by the skin. These direct dyes are the same type used in do-it-yourself dyes, which can be purchased at any supermarket or hardware store.

Other dyes used to color natural and synthetic fibers can also be harmful, but the primary danger with these is limited to skin irritation.

WHAT YOU CAN DO

Because no laws require that the type of dye used must be listed on the fabric label, it is impossible to tell if the dye on any particular fiber may be carcinogenic or may simply cause an allergic reaction. The best you can do is try to avoid dyes that bleed—if the dye comes out in water, it can be released by perspiration and absorbed by your skin.

"Mercerized" cotton fabrics have undergone special nontoxic processing to make them colorfast, and most domestic fabrics also keep their color well. But be cautious about buying inexpensive cottons from foreign countries, as these dyes are especially unstable.

In general, light-colored fabrics are safer than dark colors. Always wash fabrics before wearing to remove any excess dyes.

In my best-of-all-possible nontoxic worlds, all fabrics would be made of natural fibers and colored with natural dyes. If you are a do-it-yourselfer type and would like to try a new project, there are many natural dyes available by mail, and unfinished natural-fiber fabrics and yarns appropriate for dyeing.

NO-IRON BED LINENS AND
PERMANENT-PRESS CLOTHING ☞ 1

No-iron bed linens, permanent-press clothing, and all polyester/cotton fabrics are treated with **formaldehyde resin** finishes to keep them wrinkle-free. During the manufacturing process, the formaldehyde resin finish is applied in such a way that it becomes a permanent and irremovable part of the fiber, and it continues to release formaldehyde fumes for the life of the fabric (yes, it could go on for *years*). Newly processed textile products often release formaldehyde fumes at levels of 800 to 1000 parts per million.

Symptoms from formaldehyde vapor inhalation can include *tiredness, insomnia,* headaches, respiratory problems, coughing, watery eyes, excessive thirst, and many other common symptoms. Exposure can also aggravate asthma attacks. Skin contact with formaldehyde finishes can result in mild to severe skin rash. Remember, we spend one-third of our lives in bed with our nose next to the pillowcase and our skin rubbing against the sheets; perhaps the formaldehyde resin might explain mysterious nighttime symptoms.

Fabrics treated with formaldehyde resins are not required by law to be labeled as such; however, they will usually be easy to spot. Look for the terms *crease resistant, permanent pressed, durable pressed, no-iron, shrinkproof, stretchproof, water repellant, waterproof,* or *permanently pleated* on the label. Virtually all polyester/cotton blend fabrics have formaldehyde finishes, even if not stated on the label. Polyester/cotton bed linens have a particularly heavy finish to last through nightly use and frequent laundering. Formaldehyde is also used as a flameproofing agent on nylon fabrics.

WHAT YOU CAN DO

Since all polyester/cotton fabrics are treated with formaldehyde finishes, and natural fibers usually aren't, it's better to buy natural fibers whenever possible. With the exception of cotton, these are generally not treated with formaldehyde. Watch for labels that read "easy care 100% cotton" or "no-iron 100% cotton" as these would also have formaldehyde finishes. I have noticed that the term *no-iron* is now being used used on cotton flannel sheets as well, but don't be concerned about this. I have purchased several sets with this labeling and have found them to be free of finishes. It's true, they are naturally no-iron; cotton flannel doesn't wrinkle because of the weave of the fabric, not because it has been treated.

Always wash textile products before you use them to remove any sizing or finishes. Washing can lower the level of formaldehyde emissions from no-iron finishes from nearly 1000 parts per million when new to around 100 parts per million, but low levels of formaldehyde will continue to be released as the resin breaks down through washing, ironing, and wear, so it is best to avoid these finishes altogether, if possible.

LAUNDRY DETERGENT

From the Manufacturers' Warning Labels:

> **DANGER: In case of eye contact, get prompt medical attention. Keep out of reach of children.**
> **WARNING: Harmful if swallowed, irritating to eyes and skin. Keep out of reach of children.**
> **CAUTION: May be harmful if swallowed.**
> **(Some detergents with similar formulas have no warning on the label.)**

Detergents were developed especially to clean synthetic fibers. Detergents cause more poisonings than any other household product, most often when children accidentally ingest colorful powders packaged in easy-to-open boxes.

A lesser hazard comes from the residues left on clothing and bed sheets that can cause severe skin rashes, or from "springtime fresh" fragrances that linger long after articles are laundered. There have also been reports of flu and asthma symptoms associated with breathing air that contains detergent dust.

WHAT YOU CAN DO

Instead of using a detergent-based cleaner, try a product made from natural soap. There is a big difference between these two cleaning agents. Detergents are formulated from petrochemicals and may contain bleaches, synthetic whiteners, and artificial fragrances. Some detergents have been banned because of the environmental hazards they create. Soap, on the other hand, is made from natural minerals and fats and has been used with no ill effects for hundreds of years. Natural soap flakes are available at many supermarkets and natural food stores, or you can grate bar soap to use in the laundry. One problem with soap that our grandmothers used was that it can leave a residual scum on fabrics. But there is a solution to this problem. Keep reading . . . I'll get to it in the next entry, under the heading "Chlorine Bleach."

You don't always need to use soap to get clothes clean. Often the purpose of laundering is not so much to dispel dirt, as to freshen clothing and remove perspiration and odors. For this you can use about 1 cup of plain baking soda, white vinegar, or borax (available in the cleaning products section of your supermarket), or 1 tablespoon trisodium phosphate (sold at hardware stores) per washerload of clothes. All these natural substances are excellent for removing smells and they don't add any odors of their own.

CHLORINE BLEACH

From the Manufacturers' Warning Labels:

> **CAUTION: Keep out of reach of children. May be harmful if swallowed or may cause severe eye irritation. Never mix chlorine bleach with cleaning products containing ammonia, or with vinegar. The resulting chloramine fumes are deadly. It also should not be used on silk, wool, mohair, leather, spandex, or on any natural fiber that is not colorfast as it can damage or discolor the fabric, and cause colors to run.**

While the greatest danger of chlorine bleach is accidental ingestion, the fumes inhaled during use also pose a lesser hazard. Product labels warn only against drinking liquid bleach, but toxicology books report that chlorine is "toxic as a [skin] irritant and by inhalation." Even though the amount of fumes released is well within recognized safety limits, many people report adverse reactions to chlorine fumes at these low levels, and even to chlorine residues left in fabrics after laundering.

WHAT YOU CAN DO

I use sodium hexametaphosphate instead of bleach. This pure mineral powder makes "your whites whiter" and "your brights brighter" by dissolving the mineral deposits and soap scum that makes fabrics look dull and dingy. You can order sodium hexametaphosphate from a chemical supply house (look in the Yellow Pages) and have it shipped right to your home. It can be expensive to buy in small quantities, so you might want to consider purchasing a large quantity and sharing with friends.

To give you some idea of how well this works, a few years ago I asked my boyfriend to not bleach his clothes with chlorine bleach because the residue left on his clothes made me sneeze every time I got near him. So he followed my washing instructions, but began to complain after a few months that his shirts weren't "white" enough (dull, dingy, soap-scum buildup). As a businessman, he needed "bright white" shirts to wear with his three-piece suits. I was experimenting with sodium hexametaphosphate at the time and gave him some. He loved it! It solved the problem with one washing and he still uses it (even though he's no longer my boyfriend!).

To use sodium hexametaphosphate, add ¼ cup to 1 cup per 5 gallons of water, depending on water hardness, in your regular wash to prevent dulling film from forming. To determine the proper amount you should use, start with ⅛ cup sodium hexametaphosphate in a washerload of water and keep adding until the water feels slippery between your fingers. Add half the amount of soap you would normally use, after adding the sodium hexametaphosphate.

To remove years of accumulated detergent film from laundry,

run the items through a whole wash cycle using twice as much sodium hexametaphosphate as you would normally use, and no soap or detergent.

As a second choice, buy a powdered nonchlorine bleach, available in the laundry section of any supermarket. This product contains sodium perborate or borax as a bleaching agent instead of chlorine. This is a big improvement over chlorine bleach, but it still contains some ingredients, such as artificial fragrance, that you may want to avoid if you are allergic.

FABRIC SOFTENER

From the Manufacturers' Warning Labels:

> **CAUTION: Keep out of reach of children.**

Fabric softeners leave a residue on fabrics to control static cling. They never really wash out, so you are constantly exposed whenever you are in contact with a fabric. Residues can be very irritating to skin, and can cause allergic reactions such as stuffy nose and watery eyes.

Fabric softeners also are usually very strongly scented. People with allergies to perfume should be particularly careful about using these products.

WHAT YOU CAN DO

Fabric softeners are one of those products that didn't even exist before synthetic fibers became popular. They were developed solely to reduce static cling that builds up on synthetic fabrics. If you wear natural fibers, you don't need to use fabric softeners at all, because there is no problem with static cling. Go through your closet, separate your natural fibers from synthetics, and at least refrain from using fabric softener in your natural-fiber laundry. If you have lots of natural/synthetic blends, try laundering them without fabric softener and see if you really need to use it.

If you find you do need a fabric softener, the safest and most convenient type is the unscented sheet variety that goes into the dryer, rather than a liquid added to the wash (that might be accidentally swallowed) or an aerosol spray that is applied to dry clothes (that could end up in your eyes or lungs).

SPRAY STARCH 👄 👁 ✍ 1

From the Manufacturers' Warning Labels:

> **CAUTION: Contents under pressure. Keep out of reach of children.**

Spray starch is not generally considered to be an especially toxic product, and the label warning reflects this, pointing out only the dangers of the aerosol can.

Spray starch may contain such toxic chemicals as **formaldehyde, phenol,** and **pentachlorophenol,** but probably the biggest danger in using a spray starch comes from the mechanics of the spray. The **aerosol spray** is used to distribute the harmful chemicals mixed with a fine cornstarch powder, which could end up in eyes and lungs. Not that cornstarch is harmful, but it is a particle, and could act as a carrier for the other, more harmful, substances.

WHAT YOU CAN DO

You can make starch by dissolving 1 tablespoon cornstarch in 1 pint cold water. That's all the commercial spray starches are—just a little cornstarch mixed with chemicals. Place in spray bottle and squirt to apply (shake before using).

I don't even use starch. Most of my clothes are cotton or silk, and I like the softness of the unstarched fabric against my skin. I have found that if I want my shirts to look "crisp," I can take freshly laundered items down to a dry cleaning establishment where I can request "press only—no starch" for a nominal fee. The high heat of their pressing machines adds a certain "hold" to the fabric that I can never achieve with my iron at home (and besides, then I don't have to iron them myself!).

DRY CLEANING 👁 �🖐 👃

Many garments, made from both natural and synthetic fibers, are labeled "dry clean only."

Dry cleaning isn't really "dry"; instead of detergent and water, items are "washed" with detergent and a solvent that removes spots and stains without being absorbed by the fiber. Dry-cleaned fabrics don't shrink or stretch, dyes don't fade or run, delicate fabrics don't tear or waterspot, and wools don't mat.

Perchloroethylene is the most popular dry-cleaning solvent in use today. Inhaling fumes from this solvent can cause cancer, liver damage, depression of the central nervous system, light-headedness, dizziness, sleepiness, nausea, loss of appetite, and disorientation.

WHAT YOU CAN DO

Do dry-cleaned items pose a hazard? Yes and no. Inhaling fumes of perchloroethylene is dangerous, but perchloroethylene is a very volatile substance that evaporates thoroughly, leaving no residue. Dry-cleaned items do go through a drying process, but many times the items are still damp when covered with their protective plastic. Studies done by the Environmental Protection Agency listed fumes from slightly damp dry-cleaned items as a common indoor air pollutant. The solution is simple: As soon as you bring home dry-cleaned items, remove the plastic covering (tie it in a knot and dispose of it in a place where babies won't find it and accidentally suffocate while playing with it) and hang in a well-ventilated area (preferably outdoors) to encourage evaporation of the solvent. This could take up to a week, and the warmer it is, the faster the solvent will dry. If you have an extra room in your home, you might want to set up a space where items that have just been dry-cleaned can be hung near a small space heater. Close the door to keep fumes out of the rest of the house, and then open windows in the extra room to ventilate the solvent fumes to the outside.

You might decide that you don't want to dry clean everything marked "dry clean only" and that's okay too. These labels are frequently sewn into items that don't really need dry cleaning; manufacturers just *recommend* dry cleaning out of fear that the consumer will complain the garment is inferior after they wash it incorrectly. Even professional dry cleaners agree that you can wash almost anything without harm—if you know how to do it.

Wash cotton garments by just throwing them into the washing machine with natural soap, borax, baking soda, and/or sodium hexametaphosphate (see section on "Chlorine Bleach"). Wash whites in hot water and colors in warm water to prevent colors from fading. About a tablespoon of white vinegar per tub of laundry will also help keep colors colorfast. Using borax in the wash will help retard the formation of mildew. Dry cottons in medium heat to prevent shrinking, and remove them from the drier as soon as they are dry to minimize wrinkles.

Linen can also be tossed in the washing machine. Pretend it's a colored cotton and follow the same instructions. Remove from the drier and press while it's still damp.

Washing silk fabrics in the washer or by hand is generally discouraged, but I have been washing my silk shirts by hand for 6 years and they've all turned out fine. Wash each item separately by swishing them around in a basin of very cold water with a bit of mild soap. Do not rub. Rinse with cold water and gently remove excess water by rolling the fabric in a towel. I let my silk garments drip dry in the shower stall and then press them with a warm (not hot), dry iron.

Hand-wash wool in cool or lukewarm water (to prevent shrinkage) with mild soap or a few tablespoons of vinegar. Sweaters and knits should be reshaped on a towel to their original size while still damp. Roll up the sweater in the towel to absorb excess moisture, then dry by hanging it over the back of a wooden chair or over a towel bar.

Instead of taking your down jacket or sleeping bag to the dry cleaner, you can wash it in your bathtub with warm water and mild soap or baking soda. Dry it in a tumble drier at a cool temperature.

SPOT REMOVER

From the Manufacturers' Warning Labels:

> **CAUTION: Eye irritant. Vapor harmful. Keep out of reach of children.**

The most commonly used solvent in spot removers is **perchloroethylene,** the same solvent used in dry cleaning. Your major risk in using products containing perchloroethylene comes from your exposure while actually using the product. Perchloroethylene fumes are carcinogenic, highly toxic, and can cause light-headedness, dizziness, sleepiness, nausea, loss of appetite, and disorientation. In large amounts, exposure to perchloroethylene fumes can be fatal.

WHAT YOU CAN DO

Spots are easiest to remove the minute they occur, so get in the habit of attacking spots when they happen, before they become stains.

One day my literary agent invited me to join her for dessert at a luncheon she was having in a fancy restaurant with an important editor from a major publishing house. I ordered an ice cream creation that was swimming in a pool of bittersweet chocolate sauce. I took one bite and the entire scoop of ice cream landed on the skirt of my red cotton corduroy dress. My agent immediately ordered a bottle of club soda and doused my dress thoroughly. It didn't all come out, but when I washed it later in plain soap and water, the chocolate stain disappeared. The stain on my dress is gone, but I still think of that day every time I pass that restaurant.

You can make your own good, all-purpose spot remover by dissolving ¼ cup borax in 2 cups cold water. Sponge it on and let it sit until it dries, or soak the fabric before washing in soap and cold water. This works well for blood, chocolate, coffee, mildew, mud, and urine.

Try this for fruit juice or tea stains: Stretch the fabric over a basin and pour boiling water over the stain.

Glycerin works well to soften some spots—grass or milk, for example. It is a by-product of soap manufacture and is available at most drugstores. Glycerin soap will also work. Try rubbing grass or milk stains with glycerin and rinsing with warm water.

Undiluted vinegar or lemon juice may also work to remove some spots.

These alternatives to using chemical spot removers are worth trying, especially since chemical spot removers are so dangerous to use.

MOTHBALLS ☞ 👁 ☞ 𝄐

From the Manufacturers' Warning Labels:

> **CAUTION:** May be harmful if swallowed. Avoid prolonged breathing of vapor or repeated contact with skin. Keep out of reach of children.

Mothballs are made from 100 percent **paradichlorobenzene,** a volatile chemical that can cause severe irritation to your nose, throat, and lungs; depression; and injury to your liver and kidneys when you are exposed to it over a long period of time. It is harmful if swallowed and because mothballs look so much like candy, they are very attractive to children. If a 2-year-old child were to accidentally eat even one mothball, he or she could develop seizures in less than an hour.

I have always found the warning label on mothballs amusing. It clearly states, "avoid prolonged breathing of vapor," yet by the very design of the mothballs, you must be constantly exposed to these fumes. The odor of mothballs hidden in a closet can permeate your entire home; certainly, mothballs in the bedroom closet can infiltrate the room and increase to very high levels if the room is not ventilated adequately. The vapors from mothballs are also absorbed by clothing and blankets, making your exposure very direct when you are around these items.

WHAT YOU CAN DO

Instead of buying paradichlorobenzene, look in natural food stores and closet shops for herbal products that act as repellants. They might contain lavender or other herbs, or cedar oil, chips, or needles. You can order these products by mail. Make sure the product is of natural origin, as you may find synthetic imitations.

It might be easier for you to find the ingredients to make your own herbal moth repellants than to buy them ready-made. Make sachets of any of the following:

- Dried lavender
- Equal parts dried rosemary and mint

- Dried tobacco
- Whole peppercorns
- Cedar chips or other wood soaked in real cedar oil

Your natural food store should carry most of these items; pet stores will have cedar chips. It may be difficult to find cotton sachet bags; if so, try cotton baby socks, or sew your own.

The moths you see flying around are not the moths that eat your woolens. Two varieties of clothes moths cause damage. They are too small to notice and are not drawn to light like other varieties. It is the larvae of these moths that eat fabric, not the moth itself.

Your best protection is to store and maintain your woolens correctly to prevent larvae from hatching. Simply wearing all your clothes at regular intervals will cause any larvae to fall off. Or, you can occasionally air items in the sun, then brush them to crush larvae and remove dried up pests. Washing and dry cleaning will kill clothes moths in all stages of development, as will pressing with a steam iron, running through a hot drier, or placing items in a warm (140°) oven for an hour. You could also place small items in the freezer for several days. When you buy new woolens, it is a good idea to put them through one of these treatments before storing them with your other items to kill any larvae that might be present at the time of purchase. This is especially important if you make a point of buying unmothproofed woolens.

Once you know your woolens are free from moths, store them properly. Place them in drawers or boxes containing natural repellants or, if you are storing them over the summer, protect them in airtight containers, such as paper packages or cardboard boxes with all edges carefully sealed with paper tape. You will have secured your woolens against the munching of moth larvae, and your home will be free of dangerous fumes.

CHAPTER 9

The Home Office

Almost every home today has an office space. Maybe you run a small business from your home or do some extra work evenings and weekends. Or perhaps your office is a desk where you pay your bills and keep your schedule organized. Regardless of its size, every office probably has a few hazardous items that could be eliminated, or at least adjusted to be more healthful.

HOME COMPUTERS

You might have seen various items in the news about the possible dangers of video display terminals (VDTs)—stories about cataracts and birth defects, but everything seems vague. I heard those stories too and set out to see what information I could find. After all, this is a book on health hazards in the home, and many, many homes now have home computers with VDTs. I have one. I've been sitting here in front of it for many hours writing this book. I couldn't possibly meet my deadlines as a writer without it, keep track of my newsletter subscribers, or keep all my information on where to buy natural products up to date.

The important thing to remember in evaluating the risk to *you* is that most available data about the health hazards of VDTs apply to workplace use of these machines and not home use. The people who are suffering from eye irritation, double vision, rashes on their faces, headaches, irritability, stress, and neck and back pains work at computers hour after hour, day after day, month after month. But in some cases home users do, too. I've been sitting here at least 8 hours a day for the past few months, and if you have an enthusiastic kid and a computer, you know how many hours they can spend at it too.

Another thing to remember is that research about health effects is incomplete and contradictory. Instead of answering questions, the research just makes us ask more. All I can do here is report the suspected dangers, so at least you are familiar with the problems of current concern.

Radiation emission is probably the most confusing issue. The FDA, the American Academy of Ophthalmology, the National Institute of Occupational Safety and Health, and the National Academy of Sciences all agree that the amount of radiation given off is too small to pose a threat to health. However, the issue of radiation emissions is far from settled. No studies have conclusively linked radiation from VDTs to health problems; however, some observed frequencies in personal computers are similar to those linked to biological effects in animals, such as changes in brain waves and birth defects.

There is also some question about how low-frequency electromagnetic radiation may affect our bodies' internal electromagnetic fields. And then there's extremely low-frequency radiation, which has only recently been discovered to cause birth defects in test animals.

If you're wondering how the radiation from a VDT differs from that emitted by a television set, the difference is distance, not type. Most of us sit within 18 inches of our VDTs and across the room from our television sets. Since radiation diminishes by the square of the distance, a few feet can make a big difference in the danger.

Beyond the radiation issue, there is also a problem with static electricity, generated by the high voltages in the air that surround VDTs and the bodies of the people sitting at the keyboard. The positively charged field around a VDT neutralizes the negative ions in the air space, creating an area that is high in positive ions. High concentrations of positive ions have been associated with fatigue, metabolic disorders, irritability, headaches, and respiratory problems. In addition, dust, tobacco smoke, and chemical pollutants become positively charged and seek out the nearest grounded or oppositely charged surface, which is usually the operator's face. The particles clinging to your face can cause rashes, itchy eyes, and dry skin. I remember when I first used a computer it was in a small, unventilated office. After about an hour or so, my face would be bright red—I looked and felt like I had a sunburn.

It seems that many of the eye problems associated with VDT use are caused by glare from improper lighting. Bodily aches and pains are usually attributed to poor posture and improper furniture.

WHAT YOU CAN DO

Even though the causes of these health problems are hard to pinpoint exactly, there are some things you can do to make your computer time more healthful.

To eliminate the static electricity problem, the obvious solution is to buy a negative ion generator, a small device that emits negatively charged ions into the air. But there can be problems with that, too. The negative ion generator must be of good quality and proper design, and it must be positioned properly with respect to the VDT, the operator, and the air flow in the room. The optimal placement could be overhead, or nearby on a table. It is important not to generate *too many* negative ions, not because the negative ions have side effects, but because there is the possibility that your body will absorb negative ions faster than they can be discharged, resulting in a buildup of negative ions on your skin. When this happens, you will begin to attract positive ions, which is just what you don't want. If you want to experiment with this, try a variable-output negative ion generator and place it in different locations. You should be able to feel the difference as you search for its optimal spot. Leafy houseplants and open windows can also contribute to reducing the static electricity problem.

For glare control, indirect, diffused lighting is best. Don't let sunlight or bare bulbs shine directly on your screen. If you need brighter lighting to be able to read text you are inputting, use a small work light that shines on the paper, but not on the screen. Use the minimum contrast setting at which you can comfortably see the text; you may be able to work longer with less eyestrain at lower settings.

Place your computer equipment on a table that allows you to adjust the screen and keyboard to proper heights to fit your body. Sit in a comfortable position and then adjust the screen so you are looking slightly down at it. To adjust the keyboard height, sit in a normal position and bend your arms at the elbow to form right angles. Your hand should just fall down from there comfortably onto the keyboard.

Prevent bodily aches and pains by taking frequent breaks. Sitting motionless for long periods of time slows circulation, reduces muscle tone, and causes fatigue. Take a break every 90 minutes to stretch, walk around, and, if you can, do some other type of work for a short period of time. Some European countries are considering legislation requiring limits of 4 or 5 hours a day for work in front of a VDT. I've gotten in the habit of hitting my "save" key every few paragraphs when I'm writing. Not only is my text protected in case of a power failure, but it gives me a few seconds to wiggle in my chair.

TYPEWRITER CORRECTION FLUID

From the Manufacturers' Warning Labels:

> **WARNING: Intentional misuse by deliberately concentrating and inhaling the contents can be harmful or fatal. Nonflammable and nonhazardous when used as directed.**

Typewriter correction fluids may contain **cresol, ethanol, trichloroethylene,** and **naphthalene,** all toxic chemicals that can be fatal in high doses.

It has always been amusing to me that the label warning states "nonhazardous when used as directed." Here are the directions: "Shake well. Touch on. Do not brush. Apply sparingly. Allow 8–10 seconds to dry." Considering the toxicity of the ingredients, more appropriate directions might instruct you to "Open all the windows, turn on an exhaust fan, wear a gas mask, and use an eight-foot robotic arm." I know I'm exaggerating, but they might at least instruct you to use it in a well-ventilated area and not breathe the fumes directly.

Another reason you might not want to keep typewriter correction fluid around the house is that some teenagers use it to get high. A few concentrated whiffs of this stuff can get you high, but it can also cause respiratory arrest, cardiac arrest, and death.

HOW TO PROTECT YOURSELF

If you do a lot of typing, and make a lot of errors (as I do), it's well worth the investment to have a self-correcting typewriter or a word processor. They aren't as expensive as they used to be, so it may be more appropriate for your budget than you think (they're also tax deductible if you have a home business).

Water-based typewriter correction fluids are available in most office supply stores. They are the type intended for use with copies, rather than typewriters, but they can be used quite effectively on top copies. One coat is rather transparent, but if you let it dry and apply a second coat, it will cover up your error.

Instead of using a liquid, try the "white-out" tapes that strike a white powder over the error. If you are typing a master that will be printed or copied, you can use adhesive correction tapes to cover whole lines or paragraphs.

EPOXY, RUBBER CEMENT, AND "SUPER GLUE"

From the Manufacturers' Warning Labels:

> **DANGER: Extremely flammable. Vapor harmful. Harmful or fatal if swallowed. Skin and eye irritant. Keep out of reach of children. Bonds skin instantly. Toxic.**
> **CAUTION: Do not use near sparks or flame. Do not breathe vapors. Use in well-ventilated room. Keep away from small children.**

Adhesives are full of volatile chemicals. **Naphthalene, phenol, ethanol, vinyl chloride, formaldehyde, acrylonitrile,** and **epoxy** are a few of the chemicals more commonly used. All these substances release toxic vapors, although you probably wouldn't die from breathing them, and, in addition, some are known to cause cancer, birth defects, and genetic changes.

These products also present a secondary danger. Some of the ingredients, such as **phenol,** are very easily absorbed through the skin, an exposure we don't usually think about and yet one that is quite likely when applying glues. Our exposure to these chemicals may be small, but we shouldn't underestimate their peril. One book on toxins describes the skin destruction caused by phenol as follows: "[Phenol] exerts a powerful corrosive action that kills skin tissues with which they come in contact. The top part of the skin then becomes whitish, and the bottom part becomes reddish because of hemorrhaging. Death can result."

WHAT YOU CAN DO

The safest glues on the market are white glues and yellow woodworking glues, both widely available at most stores. White glue effectively bonds paper, cloth, wood, pottery, and most other porous and semiporous materials. It is quick-drying, clear, and "nontoxic" (as defined in the Federal Hazardous Substances Act). These glues work for many jobs you wouldn't typically use them for, such as laying hardwood floors, so when you think "glue," try white or yellow glue first.

It's probably not worth the trouble to make glue at home since you can use white glue for so many things. But if you are the do-it-yourself type and want to try something different (or if you're out of glue) here's a simple recipe for paper paste:

- Blend 3 tablespoons cornstarch and 4 tablespoons cold water to make a smooth paste. Boil 2 cups water and stir in paste (rather like thickening gravy with a flour paste). Continue to stir until mixture becomes translucent. Cool and use.

If a toxic adhesive is the *only* glue that will do the job, use in a well-ventilated area (outdoors would be best) and wear protective mask and gloves. Once completely dry they are safe, but you want to protect yourself well during application.

PERMANENT-INK PENS AND MARKERS

I know this seems like such a small thing, but I have to mention permanent-ink pens and markers for two reasons. First, when I was a little girl, my mother always wrote with fluorescent-pink ballpoint pens, and whenever I would read a note she had written, I would get sick to my stomach. Many years later, I realized that the ink in all ballpoint pens made me feel nauseated. Second, I have had several graphic artists as consulting clients who almost gave up their careers because of the symptoms they were experiencing from the solvents in colored markers. And I have known many, many people who are unable to write intelligible sentences with a ballpoint pen, but were quite literary when writing with a pencil. I have even seen people's handwriting change! I know this sounds farfetched, but if your child is having trouble writing, change his or her writing instrument, and see if it helps.

Since the inks used in pens and markers contain **acetone, cresol, ethanol, phenol, toluene,** and **xylene,** I'm not surprised at the reactions we've experienced. But I don't know of any studies that have researched the health effects of pens and markers.

WHAT YOU CAN DO

Fortunately for those graphic artists, there are several brands of water-based markers in art supply stores. They come in many colors (I think one brand has about 64 different shades) and several tip widths. Ask for water-based ink markers. They're easy to distinguish from the permanent-ink markers just by sniffing.

Every stationery and office-supply store sells ballpoint pens with water-based ink. It's not so easy to smell the difference in pens at first, but a water-based ink flows smoothly (and will smudge when wet), and a permanent-ink pen tends to get globs of ink stuck around the point (and won't smudge when wet).

CHAPTER 10

Hidden Hazards

There are some products in your home that could easily escape routine notice as hazards. They show no warning labels, and all have become so integrated into our lives that we rarely even stop to wonder if they might pose any danger.

Some of these items may be a significant source of pollution in your home, others may not even be present. I've included hazards that are known to have serious health consequences, as well as items whose toxicity is only beginning to be suspected.

POISONOUS HOUSEPLANTS

We are all aware that some plants may be poisonous, but we don't often suspect that plants we use in our homes as pleasant decorations may be harmful if they are touched or accidentally eaten by an inquisitive child.

Some plants can be deadly. Azaleas, chrysanthemums, creeping charlie, crocus, hydrangeas, lily-of-the valley, mistletoe, morning glory, oleander, and rhododendron may all cause illness requiring medical attention if eaten.

Other plants contain substances that are irritating to the skin, mouth, and tongue. In some cases, stomach upset and breathing difficulties may occur. Anthurium, Boston ivy, calla lily, dieffenbachia, philodendrum, pothos, shamrock, and spathiphyllum are examples of plants that fall into this category.

Amaryllis, buttercup, carnation, cyclamen, daffodil, daisy, ficus benjamina, geranium, holly berry, iris, poinsettia, pyracantha berry, and tulip bulb can all cause skin rashes if contact occurs. If eaten, nausea, vomiting, diarrhea, and abdominal cramps may result.

Contact with the plants themselves is not the only way a poisoning can occur. Accidental ingestion of the water in which poisonous plants have been standing can also have toxic consequences.

Even foods that we commonly eat have some poisonous parts. Avocado leaves; apple or pear seeds; rhubarb leaves; apricot, cherry, peach, and plum pits; tomato leaves; and the outer green husks of walnuts can all have toxic effects if enough is ingested. However, even a small quantity of green potato sprouts can be very toxic.

WHAT TO DO

If you have small children in the house, choose decorative plants that have no known toxic effects: African violet, asparagus fern, baby's breath, Boston fern, California poppy, camellia, coleus, dahlia, dandelion, Easter lily, forget-me-not, fuchsia, gardenia, gloxinia, grape ivy, hibiscus, impatiens, jade plant, jasmine, maidenhair fern, marigolds, orchids, pansies, peonies, petunias, roses, rubber plant, schefflera, snapdragons, spider plant, violets, wandering Jew, and zinnias. This is not a complete list; if you're not sure about certain plants, call your local poison control center for information.

If you do have toxic plants in your home, keep them out of reach of children and pets, and wash your hands thoroughly after cutting flowers.

You might want to decorate with edible plants, such as kitchen herbs or small flowers that can be used in salads. Your children may not like the taste of lobelia, nasturtiums, pansies, or violets, but it won't hurt them to nibble on these flowers and your friends will be impressed with your gourmet cooking!

GAS APPLIANCES, KEROSENE HEATERS, FIREPLACES, AND WOOD STOVES ⌐

The primary danger in using gas, kerosene, or wood to produce heat for cooking or warmth is the possibility of **carbon monoxide** poisoning. According to U.S. Consumer Product Safety Commission Product Safety Fact Sheet #13, "Each year hundreds of people die from carbon monoxide poisoning. Thousands of others suffer dizziness, nausea, and convulsions. You can't see, taste, or smell carbon monoxide. But it kills."

Carbon monoxide is produced when fuels do not burn completely. All fuel-burning appliances need air for the fuel to burn efficiently. When a generous supply of fresh air is available and the fuel is burning properly, there is little danger of poisoning. But when there is inadequate ventilation or the appliance is not operating properly, carbon monoxide is produced and can gradually overcome an unsuspecting bystander. Early symptoms of carbon monoxide poisoning are much like flu symptoms and include sleepiness, headache, dizziness, blurred vision, irritability, and an inability to concentrate. As poisoning progresses, the victim experiences nausea and vomiting, shortness of breath, convulsions, unconsciousness, and, finally, death.

Most cases of carbon monoxide poisoning in the home involve gas appliances, such as kitchen ranges, space heaters, wall heaters, central heating systems, and clothes driers. Other potential emitters are kerosene space heaters and wood- or coal-burning stoves and fireplaces.

These appliances can also emit other combustion by-product pollutants, including **formaldehyde, nitrogen dioxide, sulfur dioxide, carbon dioxide, hydrogen cyanide, nitric oxide,** and vapors from various **organic chemicals.** Even at the low levels produced from average use, possible symptoms from exposure to these byproducts include eye, nose, and throat irritation; headaches; dizziness; fatigue; decreased hearing; slight impairment of vision or brain functioning; personality changes; seizures; psychosis; heart palpitations; loss of appetite; nausea and vomiting; bronchitis; asthma attacks; and breathing problems. People with emphysema, asthma, angina, or chemical sensitivities should be particularly cautious. Because the effects of exposure to combustion by-products can be very subtle, it is sometimes difficult to connect symptoms with exposure.

The Consumer Product Safety Commission is concerned about the long-term exposure of low levels of carbon monoxide on heart patients, the ability of nitrogen dioxide and other combustion products to cause an increase in respiratory disease in children, and the potential of formaldehyde as a human carcinogen.

WHAT YOU CAN DO

The most effective way for you to avoid combustion by-products is to use all-electric appliances—ranges, heaters, water heaters, and clothes driers. Indoor air-pollution studies show that all-electric homes have significantly lower concentrations of combustion by-products than do homes with gas appliances. Because electric appliances are more expensive to run than gas appliances, be sure to inform your local utility company that you have an all-electric home and qualify for a special lower rate.

Most electric ranges are perfectly acceptable, with the exception of those that have self-cleaning ovens (they produce carcinogenic polynuclear aromatics that are on the EPA list of priority pollutants). Also acceptable, but less efficient, are ceramic cooktops, with electric heating elements hidden beneath a smooth ceramic slab. The only drawback to electric ranges is that some new models have a distinct odor that can take a year or more to disappear. Buying a used, reconditioned range solves this problem, and is less expensive too.

If the wiring in your home is not adequate for an electric range, you can use a variety of small electric appliances such as toaster ovens, coffee makers, convection ovens, crockpots, frying pans, woks, hot plates, or rice cookers, which can be operated on the standard 110-volt wiring. These can be purchased at most hardware and department stores. Choose appliances with a minimum of plastic parts and without no-stick finishes. If you have a health problem that makes you particularly sensitive to combustion by-product pollutants, it is well worth having the rewiring done for an electric range or using small appliances, despite their inconvenience.

The easiest way to convert from gas heat is to use electric

space heaters. There are two basic types. Radiant, or "quartz" heaters produce heat with exposed quartz tubes in a metal frame. They are inexpensive to buy, energy efficient, and produce fast "spot" heating, but they have two big disadvantages. First, they get very hot, so you have to keep them away from children, pets, curtains, shag rugs, or anything else that could get burned or catch fire. Second, since they are designed primarily for spot heating, sitting next to one could give you hot feet, but leave your hands frozen. Convection heaters use a tube filled with water or oil, enclosed in a metal case, to warm the air slowly in an entire room. These are more expensive to buy and operate than radiant heaters, and take longer to produce heat, but have the distinct safety advantage of remaining significantly cooler to touch.

I first bought an oil-filled convection heater that looked like a radiator, but I found it took too long to heat the room and was very expensive to run. Now I have a little radiant "log" (that I bought on sale for $13.00) that warms my feet right away and heats up the whole room in about half an hour. It costs only pennies an hour to operate, so I can leave it on as long as I want. Sometimes the house gets so hot (just from one little heater), I have to turn the heater off and open the window. I don't have children, pets, shag carpets, or drapes, so I'm not worried about the fire hazard.

Other acceptable heating methods are central forced-air electric, solar, and steam heat.

If you are sensitive to chemicals or have respiratory or heart problems, I cannot emphasize enough the importance of trying to replace your gas appliances and kerosene heaters. In my consulting practice I have seen many clients do everything I recommend except remove the gas from their homes and their health problems remain. But almost as soon as they turn off the gas appliances, they start feeling better. It's amazing.

You can protect yourself somewhat from being poisoned by combustion by-products and still use gas appliances by taking the following precautions:

- Dilute pollutants with ventilation—open windows, flues, fans, and vents. Gases from combustion by-products concentrate initially in the area around the appliance and then spread to other areas of your home as the air circulates, so catch them at the source. During cooking, for example, a hood fan can remove up to 70 percent of pollutants produced.

- Check frequently to make sure your gas appliances are functioning properly. Clean clogged stove burners and blocked flues, fix cracks and leaks in pipes, and keep up on any maintainance suggested by the manufacturer. A poorly adjusted gas stove can give off 30 times the carbon monoxide of a well-adjusted stove.

- Make sure you are using your appliances according to instructions.

- If possible, put your gas appliances in a space outside of the living area, venting the fumes to the outside and placing a tight seal between the appliances and the living space to prevent gases from spreading throughout your home.

- Use a new-model gas stove with low-heat-input gas pilot light and nongas ignition systems, which produce significantly fewer pollutants than do older stoves with pilot lights.

- Do not use a gas range or oven for heating a room.

In addition to the combustion by-products produced by gas, burning wood in wood stoves and fireplaces also produces other toxic by-products such as carcinogenic benzo(a)pyrene. It is easier to tell if these pollutants are present in your living space because they are extremely irritating to eyes, nose, throat, and lungs. Don't worry, I'm the last person who will tell you not to use your fireplace. Living in a forest by the ocean, my fireplace is burning almost constantly, all year round. However, because little research has been done on the effects of wood combustion by-products on indoor air pollution, take these precautions if you burn wood as your primary source of heat:

- Make sure wood stoves and fireplaces are installed and fitted properly. Be certain that your fireplace was constructed to be used as a fireplace and is not there just for decoration. Have it inspected to make sure it has all necessary linings and clearances and that the flue is open.

- Always keep the damper opened properly while the fuel is burning.

- Have your chimney inspected for creosote buildup when the weather starts getting cold, or periodically throughout the year if you use your fireplace often. Creosote buildup can cause flue fires and may also block the chimney, preventing escape of toxic fumes.

- Leave a window open a crack to allow pollutants to escape.
- Fix cracks or leaks in the stovepipe and keep a regular maintenance schedule to keep the chimney and stovepipe clean and unblocked.
- Guard against negative air pressure indoors and watch for downdrafts, which will push pollutants into your living space instead of carrying them up the flue.

ASBESTOS

Asbestos is a mineral fiber found in rocks that has been shown to cause cancer of the lung and stomach. Not everyone who is exposed to asbestos develops cancer, but there is no level of asbestos exposure known to be safe.

The danger of asbestos comes from exposure to tiny fibers that are inhaled and become lodged in lung tissue. These fibers are so small that you cannot see them, and they pass right through the filters of vacuum cleaners.

Not all products containing asbestos pose a health risk. The risk exists only when the fibers are released.

WHAT YOU CAN DO

In most cases, the best thing you can do is to leave the material containing asbestos just the way it is.

Here is a list of some common household products that may contain asbestos, and how you can best protect yourself from their possible hazards:

- Vinyl floor tiles and vinyl sheet flooring—There is some controversy as to whether or not asbestos fibers are released from asbestos-vinyl flooring materials during normal use. *Consumer Reports* magazine says no, but a team of French scientists found that heavily trafficked floors released significant amounts of fibers. Fibers can also be released if the tiles are sanded or seriously damaged, if the backing on the sheet flooring is dry-scraped or sanded, if the tiles are severely worn or cut to fit into place, if you use an upright carpet-sweeper type vacuum cleaner on the flooring, or if you sweep the flooring with a stiff broom.

- Patching compounds and textured paints—The Consumer Product Safety Commission banned the use of asbestos in patching compounds in 1977. According to the CPSC, asbestos is no longer added to textured paints. Older homes, however, may contain either of these materials and sanding or scraping will release asbestos fibers. If walls seem in good condition, leave them alone.

- Ceilings—Some large buildings and private homes built or remodeled between 1945 and 1978 may contain a crumbly, asbestos-containing material that has been either sprayed or troweled onto the ceiling or walls. If walls and ceilings are undamaged, leave them alone.

- Stove and furnace insulation—Asbestos-containing cement sheets and insulation may have been installed around wood-burning stoves, or oil, coal, or wood furnaces. If the insulation on or around your stove or furnace is in good condition, it is best to leave it alone. If the insulation is in poor condition, or pieces are breaking off, have it repaired or removed. The CPSC emphasizes that children should not play in an area containing insulation dust.

- Pipe insulation—Hot water and steam pipes in homes built or repaired between 1920 and 1972 may be covered with an asbestos-containing material, or pipes may be wrapped in an asbestos "blanket" or asbestos paper tape. Asbestos insulation has also been used on furnace ducts. If you have damaged insulation around pipes or ducts, leave the insulation in place and repair the protective covering.

- Wall and ceiling insulation—Homes constructed between 1930 and 1950 may contain insulation made with asbestos. Because this insulation is "sandwiched" behind plaster walls, you will not be exposed unless walls are torn down for remodeling. In this case, a contractor experienced with asbestos insulation should be used.

Not every brand of the types of materials just mentioned will contain asbestos, but some do. If you suspect there is a problem, try to determine from the label, from the installer, or from the manufacturer whether or not the material contains asbestos. Some plumbers and contractors who have experience working with asbestos products may be able to identify the products for you, or you can have them sent out for laboratory analysis.

See the Resources section at the end of this book for information on how you can find out more information about asbestos.

PLASTICS

We use plastics in virtually every area of our lives. A large group of moldable synthetic materials made from petroleum or coal, plastics can be as different from each other as potatoes are different from cucumbers and different from lettuce, yet they are all vegetables. They can be found as hard or soft solid forms, liquids that dry to solid coatings and finishes, adhesives, rigid or flexible foams, sheets, films, fibers, and filaments.

We are exposed to plastics in many different ways. There may be plastic vapors in the air we breathe, the plastic in clothing may rub against our skin, plastic may be absorbed into our food from packaging and storage containers, plastic may be in the water we drink from plastic pipes or storage bottles.

It is important to get to know different kinds of plastics and how to recognize them, because some plastics can be harmful to our health while others are not. When they were first introduced, we had no reason to even question their safety. But over the years, we are finding in one case after the other they are not as safe as we once thought. Some types of plastic are carcinogenic, others can migrate from packaging into our food and water, and still others can cause skin rashes—perhaps a minor complaint, but still a sign from our bodies that something is wrong.

Exposure to plastics is starting to be linked to many kinds of interesting effects. One prominent health researcher has observed several cases in which a polyester-covered polyurethane foam mattress was a contributing factor in sexual impotence! Both polyester and polyurethane foam are plastics, and with an additional dose of polyester and urea-formaldehyde plastic resin in the bed sheets, it's not surprising that one's body might have difficulty responding.

There are basically two types of plastics: thermoplastics and thermosets. The difference between the two lies in their chemical structure. Both are made from chains of basic molecules, but the thermoplastic chains remain detached and separable, while the molecules in the thermoset plastics are tightly bonded in weblike structures. This distinction is very important because it plays a key role in the relative safety of each kind of plastic.

Thermoplastics become soft when heated and harden when cooled, no matter how often the process is repeated. Think of these plastics as being like butter—it melts when heated, but becomes solid again when cooled. When butter is put into a mold and cooled, you have a stick of butter, otherwise it just hardens in any shape it happens to flow into.

Most thermoplastics are easy to recognize because they are soft, bendable, and have a certain amount of flexibility because of their loose chemical structure. While this is an advantage for performance, it is a disadvantage for our health because these soft plastics tend to vaporize plastic molecules, especially when they are heated. Have you ever noticed the smell of the interior of your car on a hot day, or when new synthetic carpets are installed, or when you open a plastic food container? These are all plastics vaporizing. You may notice after you have owned a plastic item for a while that it doesn't smell quite as much, but these plastics do continue to vaporize for the life of the product. As long as a thermoplastic is in existence, it vaporizes.

Each type of thermoplastic has different health effects. Some of the dangers are well documented, others are unknown and are only beginning to be suspected. Let's take a look at some of the more common thermoplastics you are likely to encounter on a daily basis, their health effects, and some examples of products made from these plastics.

Vinyl plastics, including **vinyl chloride** and **polyvinyl chloride (PVC),** are the most dangerous plastics. Products made from polyvinyl chloride/vinyl chloride include adhesives, artificial "grass," baby pants, containers for toiletries, cosmetics, household chemicals, credit cards, crib bumpers, floor tiles, food packaging, garden hoses, handbags, inflatable toys such as beach balls and swimming pools, magnetic recording tape, pacifiers and teethers, paper finishes, phonograph records, playpen covers, raincoats and galoshes, shoes, shower curtains, squeeze toys, toys, umbrellas, upholstery (furniture and auto), wall coverings, and water pipes.

PVC releases vinyl chloride, which can cause cancer, birth defects, genetic changes, indigestion, chronic bronchitis, ulcers, skin diseases, deafness, vision failure, and liver dysfunction. Some PVCs contain added plasticizers that make the plastic even more unstable. Because plasticizers are not bound to the plastic chemically they can be easily removed by water, oil, or heat. NASA banned the use of PVC in space capsules because the plasticizer outgassed and condensed in optical equipment. DEHP, a substance that appears to be a cancer-causing agent, is a plasticizer frequently added to PVC.

Perhaps the most dramatic experience I have ever had of the vaporizing of plastic was when I opened a metal tin of European cookies. I had checked the label carefully for artificial additives and preservatives, but inside was a polyvinyl chloride tray to hold the delicate cookies in place during shipping (these trays are also used for chocolates). The odor was horrible, and what was worse, the cookies were inedible—they all tasted like plastic.

Acrylic plastics are made from **acrylonitrile,** a suspected human carcinogen. Products made from acrylic plastics include acrylic fibers (clothing, blankets, carpets), adhesives, contact lenses, dentures, floor waxes, food preparation equipment, "Lucite," nonwoven fabrics (carpet backing, disposable diapers, felt, filters, sanitary napkins, shoe liners), paints, paper coatings, "Plexiglas," and wood finishes. Acrylonitrile has also been known to cause breathing difficulties, vomiting, fatigue, diarrhea, nausea, weakness, headache, and fatigue.

Polyethylene is a suspected human carcinogen. Products made from polyethylene include carpet fibers, chewing gum, coffee stirrers, drinking glasses, electrical-outlet safety covers, food containers and wrappers, heat-sealed plastic packaging, kitchenware, paper coatings, plastic bags, plastic pails and garbage cans, "squeeze" bottles, swizzle sticks, and toys.

The most popular **fluorocarbon plastic** is **tetrafluoroethylene,** better known as "Teflon." Tetrafluoroethylene can be irritating to eyes, nose, and throat, and can cause breathing difficulties. Teflon is used as a nonstick coating on clothes irons, cookware, ironing board covers, plumbing, and tools.

Two common plastics are made from the styrene monomer: **polystyrene** ("Styrofoam") and **ABS (acrylonitrile/butadiene/styrene) plastic.** Vapors released by the styrene monomer can be irritating to your eyes, nose, and throat, and they can also cause dizziness and unconciousness. Products made with styrene plastics include air conditioners, automobile dashboards, building insulation panels, car and airplane models, cleaning brushes, clocks, coasters, floor polishes, flotation devices, ice buckets, insulation on soft drink bottles, kitchen and bathroom wall tile, lighting fixtures, luggage tags, paints, poker chips, serving trays, sewing machine bobbins, silverware, throwaway hot drink cups, toys, telephones, and typewriter carrying cases.

Polyester fiber and films can cause eye and respiratory tract irritation and acute skin rashes. Products made from polyester include bedding, clothing, disposable diapers, food packaging, magnetic recording tapes, nonwoven disposable filters, sanitary napkins, tampons, and upholstery.

Nylon is a **polyamide plastic.** It is generally considered safe, but skin rashes and other types of dermatitis are common reactions to contact with nylon. Products that are made with nylon filaments and fibers include artificial "grass," automotive upholstery, bristles on toothbrushes, hairbrushes, paintbrushes, clothing, fishing lines, hosiery, mascara, pen tips, rugs and carpets, surgical sutures, and tennis rackets.

Thermosets begin as soft plastics, but their shape is set by heat and cannot be altered once it is made. A thermoset acts in a way similar to cake mix—it starts out as a liquid, but once you bake it and it becomes a solid, it stays solid, regardless of temperature. When you heat a thermoset plastic to very high temperatures (higher than you would encounter in normal, everyday use), they still don't melt; like a cake, they just disintegrate.

Because the plastic molecules in thermoset plastics are so tightly bonded, it is difficult for them to vaporize. For this reason most thermoset plastics are very hard and relatively nontoxic in comparison to thermoplastics. There are, however, two thermoset plastics that pose significant hazards.

Urea-formaldehyde plastic resins, found in particleboard, plywood, building insulation, wet-strength paper (tissues, toilet paper, paper towels), and fabric finishes are known to release high levels of formaldehyde, especially when new. The outgassing of formaldehyde decreases as the product ages, but even when some products are several years old, exposure is still great enough to cause health problems. Formaldehyde is a suspected human carcinogen and has been shown to cause birth defects and genetic changes in bacteriological studies. Symptoms from inhaling formaldehyde vapors could include cough, swelling of the throat, watery eyes, breathing problems, headaches, rashes, tiredness, excessive thirst, nausea, nosebleeds, insomnia, disorientation, and asthma attacks.

Polyurethane foam, used to make cushions, mattresses, and pillows, is the other thermoset plastic that can be dangerous. Exposure to polyurethane foam can cause bronchitis, coughing, and skin and eye problems. Polyurethane foam also releases **toluene diisocyanate,** which can produce severe lung problems.

Even greater than the everyday dangers posed is the danger of plastics in the event of a fire. Flames from burning plastics spread quickly, have extremely high temperatures, and produce large amounts of dense smoke. Ironically, plastics treated to be flame-retardant will produce more smoke when forced to burn than untreated materials. Many plastics produce toxic gases when they are burned; polyvinyl chloride turns into hydrochloric acid, burning polyurethane foam and polyester release toxic unreacted toluene diisocyanate. Since about 80 percent of fire victims are harmed not by flame, but by smoke inhalation, the toxicity of the smoke plays a big factor in the harm that could result from a fire in your home. Natural materials such as wood and cotton fibers also produce toxic combustion by-products but at a much slower rate than synthetic materials, allowing more time to escape before deadly gases accumulate to lethal levels.

WHAT YOU CAN DO

Despite the long list of potentially dangerous plastics, some plastics seem to be relatively safe.

Cellulosic plastics are made primarily from cellulose fibers from wood or cotton, and have no adverse health effects that I am aware of. Examples of products made from cellulosic plastics include acetate fibers, automobile steering wheels, body and cap of pens and pencils, eyeglass frames, toothbrush handles, typewriter keys, and shoe heels.

Melamine formaldehyde plastics, such as "Melmac" dishware and "Formica" countertops also seem to be relatively inert.

Phenolic plastic resins, more commonly called "Bakelite," may release a small amount of formaldehyde when new, but this quickly dissipates. Handles of pots and pans and clothes irons are generally made from Bakelite.

The two greatest exposures to plastic in most homes are wall-to-wall carpeting and beds. These are not brief exposures, but are with us constantly. You are literally sleeping every night in a cloud of plastic.

Instead of a synthetic mattress, you can buy a cotton futon or a cotton innerspring mattress. You may have a futon shop in your local community, or you may be able to find a local mattress factory to custom make a natural-fiber mattress for you. If not, both can be ordered by mail. They are no more expensive than a top-of-the-line synthetic mattress.

As for the wall-to-wall carpets, in my experience, these are one of the major sources of indoor pollution in the home. I know you don't really want to hear that because it is a great expense to replace them. Hardwood floors with a nontoxic finish are the best alternative.

For the most part, you *can* live without plastic. I still have a plastic television, plastic video cassette recorder, plastic electronic typewriter, plastic telephone, and plastic telephone answering machine because there are currently no alternatives available.

Plastics have only been popular since World War II. Before that, everything was made from natural materials. There are many,

many items still made from natural materials: wool diaper covers, wooden boxes, straw baskets, cotton shower curtains, glass jars, paper bags, leather shoes, wooden toys, glass dishes, wood and glass clocks, to name just a few. So keep your eyes open, and when you have a choice, look for a natural material.

You might want to start your plastic cleanup by just looking around and taking inventory of all the plastics in your home. See what is easy to replace, and what can wait. Maybe just start by packing your kids' sandwiches in foil or wax paper instead of plastic bags. One by one, you'll find replacements.

PARTICLEBOARD AND UREA-FORMALDEHYDE FOAM INSULATION 👁 ✋ 👃

Particleboard is made by pressing small wood shavings together with **urea-formaldehyde resin.** It is easy to recognize because you can see the pressed-together shavings on all sides. It is used extensively in the construction of inexpensive home furnishings, kitchen and bathroom cabinets, and in new home construction (particularly as subflooring and in doors). While particleboard itself is simple to recognize, it is often hidden behind a thin wood veneer, so check carefully! Good places to look for exposed particleboard are inside cabinets, at the ends of shelves, in corners, and in drilled holes. Most wood items that are described as "veneered" or having a "genuine oak veneer" are generally filled with particleboard.

All products made with particleboard will release small quantities of formaldehyde. Formaldehyde emissions are greatest when the product is new and decrease with time, but it takes many years for the formaldehyde to evaporate entirely.

Formaldehyde is a suspected human carcinogen, having been found to cause cancer in laboratory animals. Further, the National Academy of Sciences estimates that 10 to 20 percent of the general population may be susceptible to the following irritating symptoms from exposure to formaldehyde at extremely low concentrations:

cough, swelling and irritation of the throat, watery eyes, headaches, rashes, excessive thirst, nausea, nosebleeds, disorientation, and many other common symptoms.

While there are no warning labels on products made with particleboard, there is a warning label required on sheets of particleboard purchased at a lumberyard: "WARNING: This product is manufactured with a urea-formaldehyde resin and will release small quantities of formaldehyde. Formaldehyde levels in indoor air can cause temporary eye and respiratory irritation and may aggravate respiratory conditions or allergies. Ventilation will reduce indoor formaldehyde levels."

Another common product made with urea-formaldehyde resin is urea-formaldehyde foam insulation (UFFI). The Consumer Product Safety Commission banned use of UFFI in residences and schools in 1982, after receiving numerous complaints that exposure to this insulation caused respiratory problems, dizziness, nausea, and eye and throat irritations, ranging from short-term discomfort to serious adverse health effects and hospitalization. Despite the fact that this ban was later overturned by the U.S. Court of Appeals, the CPSC continues to warn consumers that evidence exists to indicate substantial risk is associated with use of this product.

WHAT YOU CAN DO

Instead of particleboard, choose wood items made from "solid wood" whenever possible. Even plywood, although also made with a formaldehyde resin, is preferable to particleboard.

If you are insulating your home, choose any other insulation rather than UFFI. In general, fiberglass batts are the least toxic. There is some concern that fiberglass fibers may pose the same hazard as asbestos fibers, but I know of no evidence that suggests that fiberglass inside walls might be released into your living space.

Some of the formaldehyde fumes from particleboard or UFFI can be sealed in; any kind of finish will help to some degree, although a sealant designed to be a vapor barrier will be the most effective. Ask your local hardware store to recommend an appropriate finish or order one by mail (see Resources). These vapor barrier sealants can reduce formaldehyde emissions up to 95 percent, but tend to break down after several years and require reapplication.

The most effective barrier for formaldehyde fumes is aluminum foil. Heavy-duty foil and foil-back paper (sold as foil vapor barrier at building supply stores) are more durable than standard cooking foil. These will, of course, give a rather space-age look to your furniture and your walls, but they work quite well inside cabinets where they are less visible and most needed to prevent fumes from building up inside the closed space. Use foil tape (available at hardware stores) to seal the edges and keep fumes from escaping.

If you do nothing else, at least open the window! As particleboard and UFFI continue to give off formaldehyde fumes, they can eventually reach very high concentrations in unventilated spaces. Try to leave a window open (even just a crack will help) in all rooms containing particleboard and throughout the house if your home is insulated with UFFI.

An appropriate air filter will also significantly lessen formaldehyde fumes, but don't expect a little air cleaner from your local department store to do the job (you know, the kind that is about a foot tall or wide, and mainly a fan with a little pad of charcoal and some sort of air-freshening fragrance). You'll need a heavy-duty machine with a lot of activated carbon or other adsorptive material especially formulated for formaldehyde removal (see Resources).

Common spider plants, amazingly, absorb formaldehyde! You'll need about 70 *Chlorophytum elatum* in one-gallon containers to purify the air in an average 1800 square foot energy-efficient home (a veritable jungle!), but don't be overwhelmed. Two or three one-gallon containers in the same room with your stereo speakers would help tremendously.

IONIZATION-TYPE SMOKE DETECTORS

All battery-powered (ionization-type) smoke detectors that contain radioactive materials may be a potential hazard. The radioactive materials are confined in metal containers designed to keep exposure to radioactive materials to a minimum, but since these alarms have only been in use for a short period of time, the real hazards are unknown. Any exposure to radioactive materials is suspected of creating potential adverse health effects.

WHAT YOU CAN DO

There are two types of smoke alarms on the market: ionization-type, which detects both visible and invisible signs of fire, and photoelectric detectors, which respond only to visible signs of combustion.

Interestingly enough, *Consumer Reports* magazine found that the photoelectric type detectors overall are much more sensitive and effective than the ionization type! In this case, the product that is the best buy is also the best for your health.

If you already have battery-powered smoke alarms and would like to switch, return your old smoke alarms to the manufacturer for proper disposal. The radioactive materials used in smoke alarms last for thousands of years and similar materials produced in nuclear reactors must be buried deep underground.

ARTIFICIAL LIGHT

Most of the information about the biological effects of light has been accumulated only recently. There has been much observation on this subject, but relatively few scientific studies have been conducted. The observations, however, have been interesting enough to call the safety of artificial light into question and to prompt scientific investigation.

The problem with artificial light sources seems to be that the spectrum of the light produced is not the same as the spectrum produced by natural sunlight. Over centuries our bodies have adapted to need the natural light of the sun, and when forced to spend most of our hours under unnatural light, our bodies tend to function somewhat less than optimally.

Exposure to artificial light has been associated with decreased calcium absorption (making one more susceptible to osteoporosis), dental caries, fatigue, decreased visual acuity, hyperactivity, and

186

changes in heart rate, blood pressure, electrical brain-wave patterns, hormonal secretions, and the body's natural cyclical rhythms. The National Institute of Mental Health has preliminary data that suggest a link between mood and light, with a striking relationship between light and depression.

Fluorescent lights are particularly suspect. Not only is the spectrum incorrect, but the fixtures produce an audible hum that has been connected to increased stress. Fluorescent lights also have a very fast flicker, which produces effects ranging from visual irritation to epileptic seizures.

WHAT YOU CAN DO

It seems to make sense that the light sources to which we are exposed should be similar to the lighting environment in which we evolved in nature. There is also some evidence that natural light helps the body to process toxic substances, and this action could be inhibited if sufficient natural light were not available to support body functions.

What you want to do is expose your eyes to natural daylight as much as possible. This doesn't mean direct sunlight, or even being in the sun at all. Shaded light is perfectly acceptable; in fact, it's preferable. The only criterion should be that the light produced be from the sun.

Unfortunately, it doesn't work to just look out the window because the essential ultraviolet rays are blocked by glass. If you opened the window before you looked out, that would qualify. Or, you could replace your glass windows, sliding glass doors and skylights with "ultraviolet transmitting plastic." It is important to ask for ultraviolet-transmitting plastic, as not all plastics transmit ultraviolet rays. My only concern about this suggestion is, of course, the possible toxicity of the plastic. I would hesitate to replace all my windows with plastic, only to have an environment that lacks ultraviolet rays be replaced with an environment thick with plastic fumes, especially when the windows are in direct sunlight. Eyeglasses and contact lenses can also be purchased made of ultraviolet transmitting plastic.

A less expensive and easier alternative would be to get some "full-spectrum" fluorescent or incandescent bulbs. They are available at some hardware stores and by mail. It might be simpler to find blue-tinted "daylight incandescent" bulbs as a second choice. These are not full-spectrum, but are closer to the desired spectrum than cool-white bulbs.

Of course, susceptibility to the effects of light varies among individuals in the same way we respond individually to any other factor in our environment. You should choose the light source under which you feel best.

Start by observing how you feel under different sources of light. I have noticed that I feel best outdoors or under cool-white incandescent bulbs. I have tried all the full spectrum lights I can find, and still feel best with my plain old white lightbulbs.

While I don't feel it is necessary to go right out and purchase full-spectrum lights or replace all your windows, I do agree that it is important for our bodies to be exposed to natural light on a daily basis —just as important as getting proper nutrition, sleep, and exercise. Spend as much time outdoors as you can; sitting on a screened porch, under a shaded tree, or next to an open window are all fine. If you can work it into your lifestyle, set up a work space with natural light on a porch or patio, or make a point of eating meals outdoors (if you work in an office building, be sure to get outside during your lunch hour), or take breaks for walks during the day.

When you are outdoors in direct sunlight, do not wear sunglasses, because they block the light from coming into your body through your eyes. This was hard for me to do at first because I was always squinting. Then I discovered that my squinting was actually a habit and that it did not actually make the sun any less bright. I practiced relaxing my face when in the sun so I don't squint and create wrinkles, and it works. I now have no problem with the brightness of the sun, but if it's really bright or I will be out for long periods, I wear a sun visor or wide-brimmed hat.

RADON

One of the most dangerous contaminants you may find in the air in your home is radon. Some scientists believe that exposure to radon is the highest radiation danger the American public faces. It is known to cause lung cancer in high concentrations, but its health effects at lower levels of exposure are unsubstantiated and controversial. Even so, the Environmental Protection Agency has estimated that exposure to radon may be the second leading cause of lung cancer, after cigarette smoking.

Radon is a natural radioactive gas that results from the decay of radioactive materials that may be present in rocks, soil, minerals, water, or natural gas. Soil and building materials containing radioactive substances are thought to be the major sources of radon in the home. The main risk is not from radon itself, but from radon's progeny, which directly or by attaching to airborne particles may be inhaled into the lungs. Two of the progeny emit alpha particles that have the potential to inflict 10–20 times the damage to biological tissue than similar doses of radiation from X rays.

Although the health effects of low-level exposure to radon are unclear, studies have shown that indoor concentrations of radon are generally much higher than outdoor concentrations, and are especially high in tightly sealed energy-efficient homes. Because it takes 1602 years for only half of the radon atoms to disintegrate, radon concentrations tend to become higher as time goes by, rather than disappearing on their own.

The range of indoor concentrations found in homes is very broad, since concentrations of radon are affected by the amount and rate of radon being produced by the particular conditions in your home, the volume of space in which it accumulates, and the rate of air exchange. Problems with radon in homes tend to be regional as radioactive materials from which it is generated occur in particular regions, either naturally-occurring, or from having been dumped. You may have dangerously high levels of radon in your home, or none at all. It is difficult to determine the risk, since its presence cannot be detected either by your senses or by immediate symptoms. Radon is an invisible, latent, potential killer, whose victims may die many years later, without ever making the connection that radon was responsible.

WHAT YOU CAN DO

First, you should do your best to determine whether or not you even have radon in your home. The homes most likely to contain radon are those built in uranium or phosphate mining areas, are constructed with radioactive building materials, or have radioactive materials brought to the home by water or natural gas. If you live in a high-risk home and it is also energy efficient, levels of radon will be higher.

Radon gas is generally produced under the house or in the basement and then seeps into the living space through porous materials and cracks. Have a professional give your home a good inspection and block any entries through which radon can filter into your home.

Ventilation will reduce your exposure to radon. Simply opening the window will help, and will also reduce levels of other indoor pollutants.

Again, there is no clear evidence that radon in the home is particularly harmful; however, because it is suspect it would be worth determining if radon is in your home, taking precautions for protection, and keeping up on the latest research in this field.

CHAPTER 11

Tobacco Smoke 👁 ℒ

From the Manufacturers' Warning Labels:

> **SURGEON GENERAL'S WARNING:** Smoking causes lung cancer, heart disease, emphysema, and may complicate pregnancy.
> **SURGEON GENERAL'S WARNING:** Quitting smoking now greatly reduces serious risks to your health.
> **SURGEON GENERAL'S WARNING:** Smoking by pregnant women may result in fetal injury, premature birth, and low birth weight.
> **SURGEON GENERAL'S WARNING:** Cigarette smoke contains carbon monoxide.

I've saved this chapter on smoking for last, not because it's the least dangerous product in your home—in fact it poses the greatest hazard—but because I hope its use is less widespread than the other products considered here. I'm also assuming that if you are even reading this book, you are health-conscious enough not to smoke. So, instead of writing a chapter on the dangers of smoking, I am focusing on the danger of "secondhand" smoke. Fully *96 percent* of the smoke produced by a cigarette ends up polluting the air, instead of being inhaled by the smoker. And because this side-stream waste hasn't gone through the filters inside the cigarette, it contains *more than twice* the concentration of pollutants.

The dangers of cigarette smoking to the smoker are well documented, but a good deal of controversy surrounds the investigation of the health effects of side-stream smoke on nonsmokers. Common sense, however, tells us that being around secondhand cigarette smoke can't possibly be good for us. Preliminary research has already shown that nonsmokers who inhale side-stream smoke can suffer from burning eyes, nasal congestion and drainage, sore throat, cough, headache, and nausea. Side-stream smoke is a significant health hazard for people with cardiovascular disease, asthmatics and others who have difficulty breathing, and people with allergies. Studies indicate that children of smoking parents have upper respiratory infections, bronchitis, asthma, and pneumonia more frequently than do children of nonsmoking parents. If you are pregnant, it stands to reason that side-stream smoke would cause the same damage to the developing fetus as if you smoked. If you live or work with a smoker, and do not smoke yourself, you still increase your chances of developing lung cancer, respiratory infections and breathing problems, decreased blood oxygen levels, and decreased exercise tolerance.

WHAT YOU CAN DO

Since this book focuses on toxics in the home, I won't discuss the politics of protecting yourself from secondhand smoke at work, or in other public places. For more information on nonsmokers' rights and taking action, see the Resources and Suggested Reading sections at the end of this book.

If no one smokes in your home, ask guests to step outside to smoke. Some people make "No Smoking" signs and place them strategically around their houses. I have a subtle poster printed by the American Lung Association with a picture of three cats and the caption "We Don't Want You to Smoke." It blends into my decor and gets the message across clearly. Don't put out ashtrays, and learn to say "no" when people ask if they can smoke. So what if your husband's boss or your mother-in-law smokes? I don't let anyone smoke in my home, no matter who they are. If courtesy demands that you allow someone to smoke in your home, open the windows and turn on a fan as soon as it is appropriate.

If you live with a smoker, ask that person to smoke outside, or delegate a "smoking room" indoors. If he or she refuses, agree on a "smoke-*free* room" so that at least you can retreat when you need to (the bedroom would be nice).

In households where there is a smoker, you can significantly reduce your exposure to side-stream smoke by investing in a high-quality air filter. This is not going to be cheap. I'm not referring here to the kind you buy for $39.95 at the department store with a little fan in it and a little pad of charcoal. A filter that will really do the job costs between $200–$900, depending on the size of the area to be cleaned, but isn't your health worth at least the amount of money the smoker spends on his or her cigarettes? A two-pack-a-day habit adds up to about $730.00 per year. You can buy a lot of clean air for that amount of money.

The advertising for air filters is often misleading with regard to cigarette smoke. It may prove the product "clears the air" by showing before-and-after pictures of a smoke-filled room, and in this respect it would be correct. Most inexpensive filters are very good at blowing the *visible particles* away or attracting them, but that's only half the problem. What they can't do is remove the *invisible toxic gases* that are equally, if not more, harmful. These cheap filters are virtually worthless for cigarette smoke removal on a regular basis.

What is needed to effectively remove cigarette smoke is a large amount of activated carbon to adsorb chemical gases and a particulate filter (High Efficiency Particulate Arrestance [HEPA] filters are most efficient). Because personal air filtration is a relatively new field, you won't find these filters in a regular retail store. At this point, they must be ordered by mail (see Resources).

Until you can afford to get your filter, keep the windows open as much as possible. Even using a plain fan will help remove some of the smoke.

Resources

WHERE TO BUY NONTOXIC PRODUCTS

If you'd like information on specific brand names and mail order sources of nontoxic products, the best sources are my first book *Nontoxic & Natural: How to Avoid Dangerous Everyday Products and Buy or Make Safe Ones* (J. P. Tarcher, 1984), and my bimonthly newsletter, *Everything Natural*. Here's how to order:

Nontoxic & Natural (book)			$9.95
Tax for California residents only			0.64
Shipping and handling (United States)			2.00
Shipping and handling (Canada)			5.00
Shipping and handling (world)			7.50

Everything Natural (newsletter)

	1 year	2 years	3 years
United States	$18.00	$30.00	$45.00
Canada	$23.00	$35.00	$50.00
World	$33.00	$50.00	$70.00

Special Report on Water Filters, $5.00
Special Report on Air Filters, $5.00

Send a check or money order to: Everything Natural
Box 390NH
Inverness, CA, 94937
415/663-1685

I also do private consultations by phone or at your home or place of business. Call me at 415/663-1685 for information on procedures and fees.

HOUSEHOLD HAZARDOUS WASTE DISPOSAL PROGRAMS

For practical information on how to set up a Household Hazardous Waste Disposal Program in your community, I highly recommend *Household Hazardous Waste: Solving the Disposal Dilemma* by the Golden Empire Health Planning Center. It can be ordered for $15.00 from **Golden Empire Health Planning Center, 2100-21st Street, Sacramento, CA 95818, 916/731-5050.**

NONSMOKERS' RIGHTS GROUPS

Action on Smoking and Health (ASH), 2013 H Street NW, Washington, DC 20006; 202/659-4310.

American Lung Association, 1740 Broadway, New York, NY 10019; 212/245-8000.

Californians for Nonsmokers' Rights, 2054 University Avenue, Suite 500, Berkeley, CA 94704; 415/841-3032.

Group Against Smokers' Pollution (GASP), P.O. Box 632, College Park, MD 20740; 301/474-0967.

GOVERNMENT AGENCIES REGULATING CONSUMER PRODUCTS

These government agencies will not give you information on nontoxic alternatives, but they can give you detailed information on the dangers of products around your home.

U.S. Consumer Product Safety Commission (CPSC), Washington, DC 20207, 800/638-8326. In Maryland 800/492-8363. The CPSC's primary goals are to protect the public against unreasonable risks of injury associated with consumer products, to assist consumers in evaluating the comparative safety of consumer products, to develop uniform safety standards, and to promote research and investigation into the causes and prevention of product-related death, injury, and illness. The CPSC puts out free "Fact Sheets" on hazardous products, which outline the problem and give minimum suggestions for reducing risk. Foods, drugs, cosmetics, alcohol, and pesticides are exempted from the commission's authority.

Food and Drug Administration (FDA), Department of Health and Human Services, 5600 Fishers Lane, Rockville, MD 20857, 301/443-3170. The FDA enforces the Federal Food, Drug, and Cosmetics Act and related laws to ensure the purity and safety of foods, drugs, and cosmetics. Many products that fall into these categories are subject to premarket approval by this agency. They can give you information on the safety of food additives, cosmetic ingredients, and pharmaceuticals.

Agriculture Department, Food Safety and Quality Service, Independence Avenue, between 12th and 14th streets NW, Washington, DC 20250, 202/447-3473. Meat, poultry, fruits, and vegetables are inspected and approved by the Agriculture Department.

Environmental Protection Agency (EPA), 401 M Street SW, Washington, DC 20460, 202/755-0707. The EPA regulates the manufacture and use of pesticides and has jurisdiction over all air and water pollution and hazardous waste.

Bibliography

Alderman, Donna, et al. "How Adequate Are Warnings and First Aid Instructions on Consumer Product Labels: An Investigation." *Veterinary and Human Toxicology,* No. 24, February 1982.

Banik, Dr. Allen E. *Your Water and Your Health.* Rev. ed. New Canaan, CT: Keats Publishing, 1981.

Bommersbach, Jane. "The Case Against NutraSweet." *Westword,* January 1983.

Boraiko, Allen A. "Storing Up Trouble . . . Hazardous Waste." *National Geographic,* March 1985.

Brobeck, Stephen, and Averyt, Anne C. *The Product Safety Book: The Ultimate Consumer Guide to Product Hazards.* New York: E. P. Dutton, 1983.

Brown, Halina Szejnwald, et al., "The Role of Skin Absorption as a Route of Exposure for Volatile Organic Compounds in Drinking Water." *American Journal of Public Health,* May 1984.

Calabrese, Edward J., and Dorset, Michael W. *Healthy Living in an Unhealthy World.* New York: Simon & Schuster, 1984.

California State Department of Consumer Affairs. *Clean Your Room! A Compendium Describing a Wide Variety of Indoor Pollutants and Their Health Effects, and Containing Sage Advice to Both Householders and Statespersons in the Matter of Cleaning Up.* Sacramento: California State Department of Consumer Affairs, February 1982.

CIP Bulletin. "Aflatoxin and Food." Carcinogen Information Program, St. Louis, MO.

CIP Bulletin. "Broiling and Benzo(a)pyrene." Carcinogen Information Program, St. Louis, MO.

CIP Bulletin. "Insecticide Residues in Food." Carcinogen Information Program, St. Louis, MO.

CIP Bulletin. "Nitrites and Nitrosamines." Carcinogen Information Program, St. Louis, MO.

Conry, Tom. *Consumer's Guide to Cosmetics.* Garden City, NY: Anchor Press/Doubleday, 1980.

Consumer Reports. "Are Hair Dyes Safe?" August 1979.

Consumer Reports. "Menstrual Tampons and Pads." March 1978.

Consumer Reports. "Microwave Ovens." March 1981.

Consumer Reports. "The Selling of H_2O." September 1980.

Consumer Reports. "Water Filters." February 1983.

Consumer Reports. "Smoke Detectors." October 1984.

Consumers Union. *Consumer Reports Buying Guide.* Mt. Vernon, NY: 1981, 1982, and 1983.

Cronin, Etain. *Contact Dermatitis.* Edinburgh: Churchill Livingstone, 1980.

Dadd, Debra Lynn. *Nontoxic & Natural: How to Avoid Dangerous Everyday Products and Buy or Make Safe Ones.* Los Angeles: Tarcher, 1984.

Donovan, Jennifer. "When You Really Need a Dry Cleaner." *San Francisco Chronicle,* 12 March 1985.

Environmental Action. "Birth Control Blues." May/June 1985.

Fisher, Alexander A. *Contact Dermatitis.* Philadelphia: Lea & Febiger, 1973.

Foster, Douglas. "You Are What They Eat: A Glowing Report on Radioactive Waste in the Sea." *Mother Jones,* July 1981.

Freydberg, Nicholas, Ph.D., and Gortner, Willis A., Ph.D. *The Food Additives Book.* New York: Bantam Books, 1982.

Fritsch, Albert J. (ed.) *The Household Pollutants Guide.* Garden City, NY: Anchor Press/Doubleday, 1978.

Golden Empire Health Planning Center. *Household Hazardous Waste:*

Solving the Disposal Dilemma. Sacramento, CA: Golden Empire Health Planning Center, 1984.

Gosselin, R. E., et al. *Clinical Toxicology of Commercial Products.* 4th ed. Baltimore: Williams & Wilkins Co., 1976.

Hawley, G. G. (ed.) *The Condensed Chemical Dictionary.* New York: Van Nostrand Reinhold, 1981.

Heimler, Charles, and Redmond, Tim. "Food Irradiation: Will Half-Life Replace Shelf Life?" *San Francisco Bay Guardian,* 27 February 1985.

Higgenbotham, P., and Pinkham, M. E. *Mary Ellen's Best of Helpful Hints.* New York: Warner Books, 1979.

Hunter, Beatrice Trum. *Consumer Beware! Your Food and What's Been Done to It.* New York: Simon & Schuster, 1971.

Hunter, Beatrice Trum. *Beatrice Trum Hunter's Additive Book.* New Canaan, CT: Keats Publishing, 1980.

Keough, Carol. *Water Fit to Drink.* Emmaus, PA: Rodale Press, 1981.

King, Jonathan. *Troubled Water.* Emmaus, PA: Rodale Press, 1981.

King, Jonathan. "VDT: How to Prevent 'Terminal Illness.' " *Medical Self-Care,* Spring 1984.

Kleiner, Art (ed.) "The Health Hazards of Computers: A Guide to Worrying Intelligently." *Whole Earth Review,* Fall 1985.

Lecos, Chris. "Reacting to Sulfites." *FDA Consumer,* December 1985/January 1986.

Lifton, Bernice. *Bugbusters: Getting Rid of Household Pests Without Dangerous Chemicals.* New York: McGraw-Hill Paperbacks, 1985.

Lipske, Michael. *Chemical Additives in Booze.* Washington, DC: Center for Science in the Public Interest, 1982.

Makower, Joel. *Office Hazards.* Washington, DC: Tilden Press, 1981.

Mason, Jim, and Singer, Peter. *Animal Factories: An Inside Look at the Manufacturing of Food for Profit.* New York: Crown Publishers, 1980.

Miller, Roger W. "How Onions and a Baked Potato Became Sources of Botulism Poisoning." *FDA Consumer,* October 1984.

Mishell, D. M., and Daniel, R. "Current Status of Intrauterine Devices." *The New England Journal of Medicine,* April 1985.

Mott, Lawrie. *Pesticides in Food: What the Public Needs to Know.* San Francisco: Natural Resources Defense Council, 1984.

National Research Council/National Academy of Science. *Indoor Pollutants.* Washington, DC: National Academy Press, 1981.

Nussdorf, M. R., and Nussdorf, S. B. *Dress for Health.* Harrisburg, PA: Stackpole Books, 1980.

Ott, John N. *Health and Light: The Effects of Natural and Artificial Light on Man and Other Living Things.* New York: Pocket Books, 1977.

Ott, John N. *Light, Radiation, & You: How to Stay Healthy.* Greenwich, CT: Devin-Adair, 1985.

Peterson, Iver. "Pollution: The Problem of Heating with Wood." *San Francisco Chronicle,* 7 December 1983.

Pfeiffer, Guy, M.D., et al. *The Household Environment and Chronic Illness: Guidelines for Constructing and Maintaining a Less Polluted Residence.* Springfield, IL: Charles C Thomas, 1962.

Potter, Morris E., D.V.M., et al. "Unpasteurized Milk: The Hazards of a Health Fetish." *Journal of the American Medical Association,* 19 October 1984.

Regenstein, Lewis. *America the Poisoned.* Washington, DC: Acropolis Books, 1982.

Rinzler, Carol Ann. *The Consumer's Brand-Name Guide to Household Products.* New York: Lippencott and Crowell, 1980.

Roffers, Melanie. "Sweet Seductions." *Medical Self-Care,* January/February 1986.

Saifer, Phyllis, M.D., and Zellerbach, Merla. *Detox.* Los Angeles: Tarcher, 1984.

Samuels, Mike, M.D., and Bennett, Hal Zina. *Well Body, Well Earth: The Sierra Club Environmental Health Sourcebook.* San Francisco: Sierra Club Books, 1983.

Shepard, Robin. "Color Your Hair . . . Naturally!" *The Mother Earth News,* March/April 1982.

Shuping, Edward R. "Aflatoxin Paranoia: Do You Fit the Mold?" *Vegetarian Times,* Issue #46.

Sittig, Marshall. *Handbook of Toxics and Hazardous Chemicals.* Park Ridge, NJ: Noyes Publications, 1981.

Solomon, Stephen. "How Safe Is X-Rayed Food?" *American Health,* March/April 1984.

Stark, N. *The Formula Book.* New York: Avon.

Steele, Gerald L. *Exploring the World of Plastics.* Bloomington, IL: McKnight Publishing, 1977.

Swezey, Kenneth M. *Formulas, Methods, Tips & Data for Home & Workshop.* New York: Popular Science Publishing, 1969.

Tortora, P. G. *Understanding Textiles.* New York: Macmillan, 1978.

Twenty Mule Team Borax: The Magic Crystal. Los Angeles: United States Borax & Chemical Corp.

United States Consumer Product Safety Commission. "Asbestos in the Home." Washington, DC: Consumer Product Safety Commission.

United States Consumer Product Safety Commission. "Caution! Choosing and Using Your Gas Space Heater." Washington, DC: Consumer Product Safety Commission.

United States Consumer Product Safety Commission. "Guide to Fabric Flammability." Washington, DC: Consumer Product Safety Commission.

United States Consumer Product Safety Commission. "Fact Sheet #13: Carbon Monoxide." Washington, DC: Consumer Product Safety Commission.

United States Consumer Product Safety Commission. "Fact Sheet #17: Flammable Fabrics." Washington, DC: Consumer Product Safety Commission.

United States Consumer Product Safety Commission. "Fact Sheet #34: Space Heaters." Washington, DC: Consumer Product Safety Commission.

United States Consumer Product Safety Commission. "Fact Sheet #44: Fireplaces." Washington, DC: Consumer Product Safety Commission.

United States Consumer Product Safety Commission. "Fact Sheet #55: The Federal Hazardous Substances Act." Washington, DC: Consumer Product Safety Commission.

United States Consumer Product Safety Commission. "Fact Sheet #67: Oven Cleaners." Washington, DC: Consumer Product Safety Commission.

United States Consumer Product Safety Commission. "Fact Sheet #72: Drain Cleaners." Washington, DC: Consumer Product Safety Commission.

United States Consumer Product Safety Commission. "Fact Sheet #79: Furnaces." Washington, DC: Consumer Product Safety Commission.

United States Consumer Product Safety Commission. "Fact Sheet #92: Coal and Wood Burning Stoves." Washington, DC: Consumer Product Safety Commission.

United States Consumer Product Safety Commission. "Fact Sheet #97: Kerosene Heaters." Washington, DC: Consumer Product Safety Commission.

United States Department of Agriculture and United States Department of Health and Human Services. "Nutrition and Your Health: Dietary Guidelines for Americans." Home and Garden Bulletin, No. 232.

United States Department of Health, Education, and Welfare. "Everything Doesn't Cause Cancer." Washington, DC: National Cancer Institute. NIH Publication, No. 80-2039, April 1980.

United States Environmental Protection Agency. "Pesticide Safety Tips." Washington, DC: Environmental Protection Agency.

United States Food and Drug Administration. "Food Poisoning: The 'Infamous Four.' " An FDA Consumer Memo.

United States Office of the Federal Register. *Code of Federal Regulations.* Washington, DC: U.S. Government Printing Office, 1981.

United States Statutes at Large. *Safe Drinking Water Act,* Vol 88 Pt 2, Public Law 93-523. Washington, DC: U.S. Government Printing Office, 1976.

Wallace, Lance A. "Personal Exposures, Outdoor Concentrations, and Breath Levels of Toxic Air Pollutants Measured for 425 Persons in Urban, Suburban and Rural Areas." Unpublished report presented at the annual meeting of the Air Pollution Control Association, 25 June 1984, San Francisco, CA.

Warde, John. "Flower Power." *Organic Gardening,* May 1985.

Waters, Enoc P. "What About Bottled Water?" *FDA Consumer,* May 1974.

Weiss, G. (ed.) *Hazardous Chemicals Data Book.* Park Ridge, NJ: Noyes Data Corporation, 1980.

Winter, Ruth. *A Consumer Dictionary of Cosmetic Ingredients.* New York: Crown Publishers, 1976.

Winter, Ruth. *A Consumer Dictionary of Food Additives.* New York: Crown Publishers, 1978.

Woo, Olga, Pharm.D. "Your Guide for Plant Safety." San Francisco Department of Public Health.

Wylie, Harriet. *420 Ways to Clean Everything.* New York: Harmony Books, 1979.

Yiamouyiannis, John, Ph.D. *Lifesavers Guide to Fluoridation.* Delaware, OH: Safe Water Foundation, 1982.

Zamm, Alfred V. *Why Your House May Endanger Your Health.* New York: Simon & Schuster, 1980.

Zimmerman, David R. *The Essential Guide to Nonprescription Drugs.* New York: Harper & Row, 1983.

Index

PRODUCTS

HAZARDS

HEALTH EFFECTS

unconsciousness, 75, 86, 92, 103, 116, 171, 180
urinary irritation or infection, 32, 105

vaginal infections, 100
vision problems, 40, 68, 100, 136, 162, 171, 179, 186
vomiting, 25, 32, 68, 92, 97, 101, 116, 136, 170-171, 179

weakness, 40, 117, 141, 179
weight gain, 99